Historical Association Studies

The Agricultural Revolution

Historical Association Studies
General Editors: M. E. Chamberlain and James Shields

China in the Twentieth Century
Paul Bailey

The Agricultural Revolution
J. V. Beckett

The Ancien Régime
Peter Campbell

Decolonization
The Fall of the European Empires
M. E. Chamberlain

Gandhi
Anthony Copley

The Counter-Reformation
N. S. Davidson

British Radicalism and the French
Revolution 1789–1815
H. T. Dickinson

From Luddism to the First Reform Bill
Reform in England 1810–1832
J. R. Dinwiddy

Radicalism in the English
Revolution 1640–1660
F. D. Dow

Revolution and Counter-Revolution
in France 1815–1852
William Fortescue

The New Monarchy
England, 1471–1534
Anthony Goodman

The French Reformation
Mark Greengrass

Politics in the Reign of Charles II
K. H. D. Haley

Occupied France
Collaboration and Resistance 1940–1944
H. R. Kedward

Secrecy in Britain
Clive Ponting

Appeasement
Keith Robbins

Franklin D. Roosevelt
Michael Simpson

Britain's Decline
Problems and Perspectives
Alan Sked

The Cold War 1945–1965
Joseph Smith

Bismarck
Bruce Waller

The Russian Revolution 1917–1921
Beryl Williams

The Historical Association, 59a Kennington Park Road,
    London SE11 4JH

The Agricultural Revolution

J. V. BECKETT

Basil Blackwell

Copyright © J. V. Beckett 1990

First published 1990

Basil Blackwell Ltd
108 Cowley Road, Oxford, OX4 1JF, UK

Basil Blackwell, Inc.
3 Cambridge Center
Cambridge, Massachusetts 02142, USA

British Library Cataloguing in Publication Data
A CIP catalogue record for this book is available from the British Library.

Library of Congress Cataloging in Publication Data

Beckett, J. V.
      The agricultural revolution/J. V. Beckett.
          p.   cm. — (Historical Association studies)
      Includes bibliographical references.
      ISBN 0−631−16287−9
          1. Agriculture — Economic aspects — Great Britain — History.
      2. Agriculture — Great Britain — History. 3. Agricultural
      innovations — Great Britain — History. I. Title. II. Series.
      HD1923.B43   1990
      338.1′0941 − dc20

Typeset in 10 on 11 Ehrhardt by Setrite Typesetters Limited
Printed in Great Britain by Whitstable Litho Ltd., Whitstable, Kent.

Contents

Maps

Acknowledgements

The author wishes to thank Professor G. E. Mingay and Dr Michael Turner for reading and commenting on an earlier draft of this book. The faults that remain are the author's own.

Introduction

What was the Agricultural Revolution? If we loosely adapt the criteria expected of an industrial revolution we would expect it to be characterized by technological change affecting the broad range of farming and not merely one sector or another, and by the substitution of machinery for labour. In turn the effects of these changes would be to move the agricultural sector of the economy to a new level of productivity which it was able to sustain. It used to be argued that these were precisely the changes which occurred in British agriculture between about 1760 and 1830, the classic dating of the Agriculture Revolution. Parliamentary enclosure, the introduction of the seed drill, the invention of new crop rotations in which roots and artificial grasses were cultivated, and improvements in livestock breeding were together thought to have had a profound impact on arable and pastoral farming alike. This was the major argument of Lord Ernle in *English Farming Past and Present*, which was first published in 1912. Ernle's book was reprinted many times, and for half a century it remained the standard volume on agricultural development. However, one by one his arguments were undermined, and the book was finally superseded in the literature by the publication in 1966 of J. D. Chambers's and G. E. Mingay's *The Agricultural Revolution 1750–1880*. They benefited greatly from 50 years of post-Ernle research and threw considerable new light on the mechanisms of change. However, the outside dates that they employed continued to reflect a primary concern with the increase in agricultural output in the later eighteenth and nineteenth centuries.

Since the 1960s the debate has moved on, and as research has proceeded the dating of the Agricultural Revolution has become increasingly uncertain. Considerable emphasis is now placed on changes which occurred before rather than after 1750, and in this process of revision the Agricultural Revolution has been divorced from its original bedfellow the Industrial Revolution, which historians are increasingly pushing into the nineteenth century. By contrast, in agriculture the years after *c*.1820 are now viewed as witnessing little short of a second revolution.

Despite these changes in outlook and approach there is at present no single volume offering students of the Agricultural Revolution a straightforward guide through the labyrinthine byways of academic discussion. This book attempts to lay down such a pathway. The first chapter provides an overview of how historians' thinking about the Agricultural Revolution has progressed. The clarity of Ernle's vision has been obscured by the mass of evidence which has come to light since 1912. Today the current of opinion holds that change in agriculture has not occurred rapidly, but has been a long and complex process varying considerably in its impact and timing across different farming regions and terrains.

Three areas of change lay at the heart of the Agricultural Revolution: innovation, enclosure and the distribution of land. As chapters 2 to 4 show, these were far more complex than was once believed. Innovation occurred over centuries rather than decades, and what was appropriate for one area was not necessarily adaptable elsewhere. Enclosure, traditionally associated with Parliamentary legislation and rapacious landlords, is now viewed as a longer-term process affecting large areas of land whether or not an Act of Parliament was involved. The disappropriation of the owner–occupier and the small farmer, together with the rise of large, capitalist-orientated farms, was by no means as straightforward as was once believed. Chapter 5 looks at how innovation, enclosure and changes in land distribution were reflected in output. Finally, chapter 6 raises the question of how our changing view of the Agricultural Revolution has altered the way in which the relationship between agriculture and the national economy is perceived.

1 The Agricultural Revolution in History

Until quite recently, historians were in no real doubt about the Agricultural Revolution. It was a movement closely associated with the Industrial Revolution, which brought changes in the methods, organization and levels of output in farming between about 1706 and 1830. The link between the two revolutions dates from the second half of the nineteenth century when historians first became fascinated by the process of change in the English economy. In his path-breaking *Lectures on the Industrial Revolution* Arnold Toynbee wrote that 'an agrarian revolution plays as large part in the great industrial change of the end of the eighteenth century as does the revolution in manufacturing industries, to which attention is more usually directed' (1919, p. 68). As the carpet of history was rolled back, changes in techniques and in the organization of farming came to light which were believed to have radically altered the agrarian structure and increased the output from land in such a way as to support the growing towns of industrial England. All this, it was argued, added up to nothing less than an Agricultural Revolution.

The early chroniclers of the revolution saw it in terms of an *agrarian* revolution. Karl Marx thought in terms of an Agricultural Revolution in which enclosures, sheep-farming, rising prices and long leases created a class of 'capitalist farmers' and expropriated labourers by the end of the sixteenth century. Toynbee followed Marx in emphasizing the importance of enclosure and the concentration of landownership in preparing the ground for the adoption of new farming methods, but he believed the vital period was rather later, from about 1760. He saw the Agricultural Revolution as being associated with 'the destruction of the common-field system of cultivation; the enclosure, on a large scale, of common and waste lands; and the consolidation of small farms into large' (1919, p. 68).

This emphasis on agrarian structures attracted the attention of European historians who studied the process of agricultural change in England partly to try to understand the land tenure problems which were endemic in countries such as Russia and Germany. Their counterparts in England were more interested in finding an answer to the contemporary problem of the agricultural depression. The depression was characterized by a downward movement in prices which began in the early 1870s and lasted until 1896, although the long-term effects were such that English agriculture did not really recover until the Second World War. For many commentators the depression was a direct consequence of the rise of the great estate and the large farm, and they argued that the solution was to be found in smaller farms and more intensive cultivation. Bowing to pressure, in 1892 the Liberal government passed a Small Holdings Act to allow local authorities to buy land and make it available to labourers wishing to become freeholders.

The importance of small holdings, and the supposedly detrimental impact on farming of the trend over the previous century or so towards large farms and substantial tenancies, was a central theme of the supporters of the Liberal government's action. Gilbert Slater, writing several years later in 1907, and J. L. and Barbara Hammond in 1911, re-emphasized this argument. In their view, the Parliamentary enclosure movement was a massive swindle perpetrated by the greater landowners to defraud their lesser brethren. By arguing in this way they were building on an earlier romantic tradition stretching back through William Cobbett's writings of the 1820s to Oliver Goldsmith's in the eighteenth century. However, the image they suggested of a dispossessed peasantry was an evocative one, particularly in the way they used it to argue that the restoration of such a group would be a means of reviving agriculture. Land, according to the Hammonds, had been wrenched from the smaller owners and farmers at the time of enclosure, and the countryside had been deprived of the backbone of the farming system which had made England great, the rural peasant (Hammond, J. L. and B., 1911).

Other historians became interested in the technical aspects of raising output. Potentially a reorganization of tenures might do something for output, but more fundamental changes were required if this was to be realized. The link between tenures and technology was established by Ernle in 1912. Ernle was no supporter of the Liberal land policy. He argued that a number of pioneer innovators had championed large farms and the deployment of large amounts of capital, and had been responsible for transforming the English countryside — hence the Agricultural Revolution. Enclosure, which had brought with it large farms, had therefore been a good thing since it had encouraged capitalist, productive farming. The result was what he called an 'astonishing' — he was reluctant to use the word 'revolution' — change in English agriculture which coincided with the Industrial Revolution, and ensured that the new urban workforce was fed. 'Farmers of the eighteenth century,' wrote Ernle, 'lived, thought, and farmed like farmers of the thirteenth century' (1961, p. 220) and it was only from about 1760 that English agriculture burst out from its medieval constraints.

Ernle's Agricultural Revolution was characterized by two features: the traditional emphasis on the post-1760 Parliamentary enclosure movement and, what was more novel, the adoption of new crops, notably turnips and clover, new crop rotations and machinery, and improved livestock breeding. For Ernle, agricultural progress was prompted by progressive landlords, particularly those in Norfolk, who grew turnips, used the seed drill, adopted the hoe to eliminate weeds, and grew clover to increase grazing. With farmers keeping more livestock they also had more manure and this, together with a number of other developments, helped to raise output. The credit for this progress lay with a handful of energetic innovators:

The great changes which English agriculture witnessed as the eighteenth century advanced, and particularly after the accession of George III (1760), are, broadly speaking, identified with Jethro Tull, Lord Townshend, Bakewell of Dishley, Arthur Young, and Coke of Norfolk. With their names are associated the chief characteristics in the farming progress of the period, which may be summed up in the adoption of improved methods of cultivation, the introduction of new crops, the reduction of stock-breeding to a science, the provision of increased facilities of communication and of transport, and the enterprise and outlay of capitalist landlords and tenant-farmers. (Ernle, 1961, p. 149)

Pride of place, according to Ernle, lay with Jethro Tull. Tull, the inventor of a mechanical system of drilling seeds at wide intervals, was also responsible for publicizing the process of hoeing between the rows; according to Ernle, 'in the progress of scientific farming Tull is one of the most remarkable pioneers'. Of the others, Viscount Townshend was a government minister in the 1720s, but following his resignation in 1730 he retired to his country estate at Raynham in Norfolk where he devoted himself

3

'to improvements in the rotation of crops, and to the field cultivation of turnips and clover, which in the preceding half century had been successfully introduced into the country'. Robert Bakewell, whose farm was near Loughborough, was best known for his pioneering work in animal breeding at Dishley. Arthur Young was the prolific agricultural writer. Finally, Thomas William Coke of Holkham in north Norfolk 'stood at the head of the new agricultural movement'. He turned a landscape without wheat, and with only scanty yielding rye and undernourished animals, into a thriving and progressive agricultural estate (1961, pp. 217−19). Collectively, the influence of these men was far-reaching:

> The improvements which these pioneers initiated, taught, or exemplified, enabled England to meet the strain of the Napoleonic wars, to bear the burden of additional taxation, and to feed the vast centres of commercial industry which sprang up, as if by magic, at a time when food supplies could not have been provided from another country. (Ernle 1961, p. 149)

Since 1912 none of these bucolic heroes has survived close scrutiny. Jethro Tull was not the originator of the seed drill, and nor was the drill particularly well known before the 1820s (Marshall, 1929−30). Turnips were introduced into Norfolk long before Townshend's split with Sir Robert Walpole in 1730; indeed, he was only a boy when they were first grown on his own family estate (Riches, 1937). Robert Bakewell's work was limited. Arthur Young may have been a prolific author but his success as a farmer left much to be desired. However, it was the case of Coke which was the most extraordinary. Ernle seems to have used contemporary accounts without appreciating the extent to which they were inspired by Coke and partially, if not seriously, flawed (Parker, 1975).

The second pivot of Ernle's argument was Parliamentary enclosure. Marx, Toynbee, Slater and others laid great stress on the role of enclosure, but by the time Ernle wrote in 1912 doubts were already being raised about its real impact. In 1909 A. H. Johnson published a book in which he used evidence drawn from Land Tax Assessments for the period 1780−1830 to argue that in many cases the so-called peasantry had been driven from the land before enclosure, particularly on the heavy Midland claylands where the majority of early Parliamentary enclosure took place. During the French and Napoleonic war years of 1793−1815 the number of freeholders increased, and small farms continued to flourish in

4

some regions through the nineteenth century. Ernle, however, was either unaware of, or chose to ignore, the implications of these findings.

In spite of these misinterpretations, Ernle's book was a remarkable achievement given the limitation of his sources, particularly his reliance on farming books and literary sources rather than farming records. His achievement was such that for half a century his book remained *the* textbook on the development of English farming. It was reprinted and revised down to a sixth edition in 1961 when, for the first time, the weaknesses of the argument were fully exposed. Critical introductions by G. E. Fussell and O. R. Macgregor pointed to the volume of literature published since 1912, and to the various sources which Ernle had overlooked. These had shown that Ernle's link between enclosure and improvement was too simplistic, that the turnip was not the key to everybody's farming problems, and that even in Norfolk the adoption of new methods took a very long time.

The 'cows and bucolic heroes' view the Agricultural Revolution could not be sustained, particularly when it became clear that many of the innovations dated by Ernle to 1760 or later could be found perhaps a century earlier (John, 1960). In the words of one historian:

> Between the middle of the seventeenth century and the middle of the eighteenth century, English agriculture underwent a transformation in its techniques out of all proportion to the rather limited widening of its market. (Jones, 1965, p. 1)

The time was clearly ripe for a new look at the Agricultural Revolution, and this was provided by Chambers and Mingay in 1966. Ernle's book, they suggested, 'as a text must be considered seriously out of date'. However, although they still chose to regard the Agricultural Revolution as a phenomenon of the decades after 1760, nearly one quarter of the book was devoted to the period before the mid-eighteenth century because 'the ancient structure was not so backward nor so incapable of improvement as was once supposed' (1966, pp. v, 52). Perhaps more logically, Mingay has since suggested that 1700–1850 should be seen as the outside dates for the Agricultural Revolution (1977).

Few historians doubted that the major increase in agricultural output must have coincided with the growth of population and the structural shift of the labour force from the country to the town in the course of the nineteenth century. Chambers and Mingay high-

5

lighted the fact that these developments could not have occurred had the farming community been unprepared for change. Just as historians of the Industrial Revolution had been stressing the importance of changes before 1760, so Chambers and Mingay pointed to the same period as the one in which major innovations occurred which transformed agricultural output in the later eighteenth and nineteenth centuries. Gradually historians were groping towards the view that the really critical changes in agricultural practice occurred over a long period from the mid-seventeenth century if not earlier, and that the post-1760 period did not offer the major discontinuity Ernle had supposed.

While the message of Chambers's and Mingay's book was still being digested, a much more radical interpretation was offered by Eric Kerridge in 1967. Kerridge argued that pressure on resources in the sixteenth century, in the shape of land hunger and population increase (with an associated rise in demand for foodstuffs), led to the reorganization of agriculture and to improvements in farming practices which Ernle had overlooked. As a result he had produced an erroneous dating for the Agricultural Revolution which, Kerridge argued, 'took place in England in the sixteenth and seventeenth centuries and not in the eighteenth and nineteenth' (1967, p. 15). Kerridge's views have aroused considerable controversy, but the weight of his evidence supported the contention that the mid-eighteenth century was no longer acceptable as a starting date for the Agricultural Revolution. Technological change had clearly been in progress for two, and possibly three, centuries before the magic date of 1760.

Taken together with the implications of Chambers's and Mingay's book, and the variation in farming practices resulting from regional differences (Thirsk 1987; see maps 1 and 2), a new viewpoint began to emerge in which agricultural change was depicted as taking place over a long period of time with considerable local differences. Innovations adopted in one area in the sixteenth century may have taken another 200 years to reach other places. In East Anglia the spread of some improvements was evident as early as the 1580s, while in north-west England it was after 1750 that progress was most marked. These chronological variations were re-emphasized by the publication in 1985 of vol. 5 of *The Agrarian History of England and Wales*, which contained twelve chapters on regional farming systems (Thirsk, 1984–5).

To complicate the matter still further other historians were beginning to question whether agricultural progress was sufficiently advanced by *c*.1820 for the term 'revolution' to be warranted, at

Map 1 English farming countries in the age of the Agricultural Revolution.
Source: J. Thirsk, *England's Agricultural Regions and Agrarian History, 1500–1750.* (London: Macmillan, 1987).

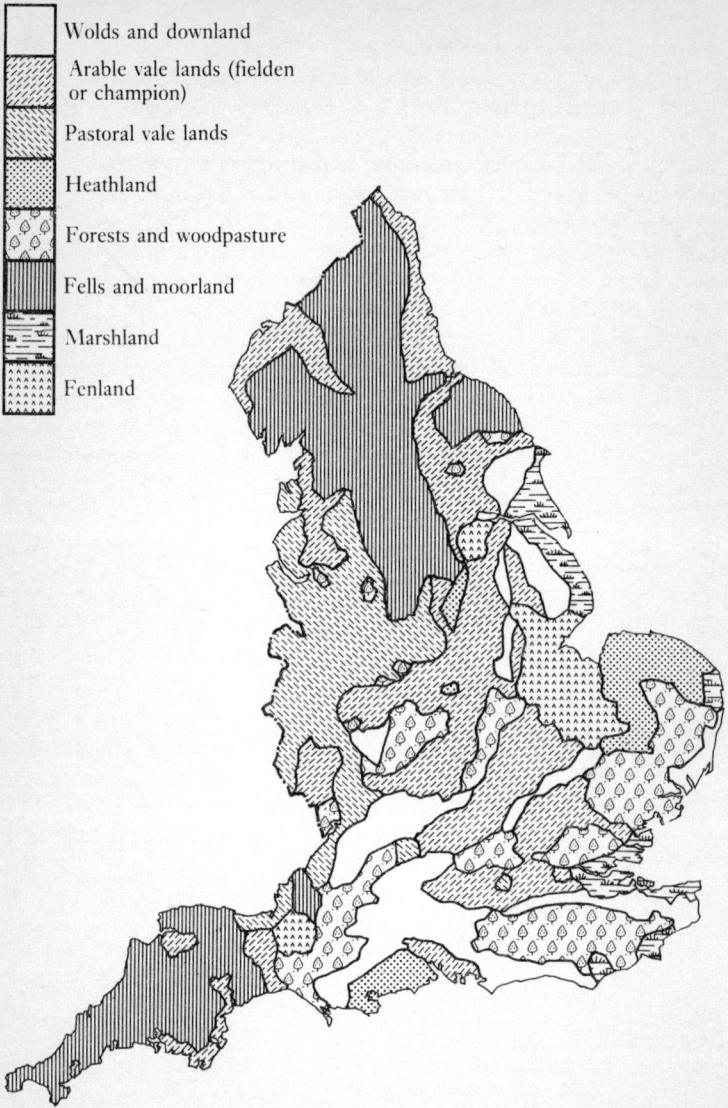

Map 2 Farming regions: a simplified schedule.
Source: J. Thirsk, *England's Agricultural Regions and Agrarian History, 1500–1750*. (London: Macmillan, 1987).

Legend:
- Wolds and downland
- Arable vale lands (fielden or champion)
- Pastoral vale lands
- Heathland
- Forests and woodpasture
- Fells and moorland
- Marshland
- Fenland

least by comparison with changes in the middle decades of the nineteenth century. Was it not the case, they argued, that the really important period of growth and development came after 1815? By then the process of enclosing the open fields, and improving rotations and livestock breeding, had reached a conclusion. Now farmers began to purchase feeding stuffs and fertilizers rather than to rely on farm produce. Coupled from the 1840s with field drainage and the construction of new farmsteads these changes seemed to suggest a qualitatively distinct new era. Moreover, the claylands, which had been unsuited to many of the new techniques developed in the seventeenth and eighteenth centuries, passed through their Agricultural Revolution only in the third quarter of the nineteenth century, partly as a result of improved drainage (Thompson, 1968).

By the 1970s historians had stretched the Agricultural Revolution(s) to cover most of the period 1560–1880. First came a period of scattered and faltering innovation in the sixteenth and seventeenth centuries. This was followed by the diffusion of the new ideas and their widespread adoption during the period 1660–1760. Finally, from 1820 or thereabouts came a series of major innovations perhaps amounting to a second revolution. But was this all so gradual as to make the term meaningless (Woodward, 1971)? E. L. Jones has written that:

> no convincing date can be found during the [eighteenth] century for any process that might be labelled an 'agricultural revolution'. The agricultural growth of the eighteenth century was, as it were, a part of the history of an expanding universe not something that began with a 'big bang'. (Jones, 1981, p. 66)

Joan Thirsk, the leading authority on English agricultural history, has even suggested abandoning the term 'Agricultural Revolution'; in her view it would make more sense if 'improvements were analysed as a continuum, to be divided between periods of more and less rapid change' (1987, pp. 57–8). After all, the spread of new agricultural crops and methods, and their incorporation within the varied farming systems which characterized the English landscape, did not occur overnight. Changes were put into effect slowly and innovations had to be adapted to the needs and opportunities of England's many farming regions.

The primary task of the farming community was to feed the population, and if this is taken as the essential criterion for defining an agricultural revolution then the concept is clearly justified. Kerridge's claim for an agricultural revolution in the period 1550−1750 was grounded on the ability of domestic agriculture to feed a population which doubled over the period. Similarly, Chambers and Mingay argued for an agricultural revolution between the mid-eighteenth century and 1880 because some 6.5 million additional people were being fed in 1850 compared with a century earlier. Although this was achieved partly from imports, it also reflected considerable increases in the output of the home-farming community. Whether we call it a revolution, or reject the title in preference for something with more gradual overtones, it is indisputable that agriculture passed through a series of remarkable changes in technology and organization which helped to raise output to much higher levels than had previously been achieved.

2 Innovation

We would expect the Agricultural Revolution to increase the output of food. This could be achieved in several ways. One was to extend the area of land in production as a means of raising output. However, all the best farming land was already in use and on poorer soils the returns were likely to be limited. Admittedly, some inaccessible areas may have been ignored in the past, and other fertile land left out of cultivation because of the labour or other resources required to clear or drain it. Thus, fen and marsh land ultimately proved enormously fertile when it was adequately drained. However, the Agricultural Revolution was not merely about extending the cultivated acreage, it was also about raising the output of the land already in use by growing new crops, by improving the methods of cultivating those already known, and by increasing the fertility of the soil by better use of manure and fertilizer. Improved strains of animals could raise the output of meat and related products, while the introduction of machinery and other labour-saving devices was a means of raising output without necessarily increasing the number of people working the land. In this chapter we shall look at what are usually seen as the vital innovations of the Agricultural Revolution, and at why they took place.

Arable Farming

Arable farming traditionally followed a set course. A varying number of grain crops was taken from the land in annual succession, and it was then left unsown – fallow – in order to recover from the years of cropping before being replanted. The pattern or rotation of cropping and fallow varied according to soil and terrain. The best known rotations were associated with open- or common-field cultivation, in which the farmers worked the village lands under the controlling eye of a manor court. Open-field farming was practised across much of the country, especially in the Midland counties, and usually involved two, three or four field rotations. In the Nottinghamshire village of Laxton, where the tradition of open-field farming is still maintained, the rotation is winter-sown wheat in one field, spring-sown barley (or beans or peas) in the second field, with the third field fallow. The cycle moves on each year, so that in any three-year period each of the fields is in wheat, barley and fallow (Beckett, 1989).

The crucial determinant of the rotations, whether or not they were in open fields, was the time needed to restore the soil after successive crops had been taken. In areas of poorer soils alternatives were practised in which parcels of land were taken in and cropped for a number of years and then allowed to go out of cultivation. Whatever the cycle, the fallow was recognized as a vital element in the system because it allowed nitrogen to be restored to the soil from the atmosphere, and permitted the repeated ploughing which was important in controlling weeds. Sheep were often run on the fallow for the same purpose. The major drawback was that one third or more of the arable land was uncropped each year.

Not surprisingly, farmers searched for ways of making better use of the land, and the technological innovation at the heart of the Agricultural Revolution was the introduction of new fodder (animal-feeding) crops which cut out the need for the fallow. Two crops were particularly important, legumes and turnips. Legumes — clover, sainfoin, trefoil and lucerne — boosted yields by raising the nitrogen content of the soil. Animals were fed on the crop which in turn improved the quality of manure returned to the land. Although the process by which this occurred was not fully understood until the nineteenth century, farmers were aware that legumes yielded heavy crops of hay — clover was a highly nutritious feed, twice as rich as ordinary native grasses — and that the fodder crops restored soil fertility after the exhausting crops of wheat and barley.

Turnips were introduced into England from Holland as a garden vegetable in the sixteenth century, but for the Agricultural Revolution they became important when they were cultivated as a field crop. Provided they were manured and hoed they helped to keep weeds under control and to provide animal feed. Turnips were also important in enabling farmers on naturally thin and infertile 'light' soils to grow arable crops on land which had not previously been cultivated. The combined effect of legumes and turnips was to reduce the fallow while increasing the output of animal feed. ers were able to keep more livestock, and this in turn increased the supply of manure. Animal waste products were the main fertilizer used on arable land, so with more manure available soil fertility improved and yields per acre rose.

This method of using root crops (turnips) and clover in conjunction with the traditional corn crops was known as alternate husbandry. Clover might be undersown with barley, and turnips were grown in the rotation between two grain crops. The most celebrated of the new rotations developed as a result of the introduction of turnips and clover was the so-called Norfolk four-course in which wheat, turnips, barley and clover followed each other in annual succession. However, farmers often wanted to grow crops such as oats for feeding horses, and other rotations were introduced over time. Progress in farming was a matter of adapting turnips, clover, and other 'artificial' grass crops such as ryegrass, cocksfoot and meadow fescue, to local conditions.

Different regions had differing requirements. In western England clover was better adapted than turnips to the conditions of husbandry. In the south-west, where grass grows through the winter months and animal fodder was never in short supply, the four-course rotation had few exponents. Similarly, it was seldom adopted in the Midlands where the clay soil was unsuited to the new crops and innovation was virtually impossible without improved drainage. Reasonably good grain crops continued to be grown, but until the introduction of underdrainage − and particularly pipe drainage from the 1840s − the claylands remained unsuited to the new crops and the traditional rotations continued.

The critical question has concerned the timing of these innovations. Ernle attributed much of the credit for their introduction to Lord Townshend:

> In 1730 Lord Townshend retired from political life to Raynham in Norfolk. There he devoted himself to the care of his estates, experimenting in the farming practices which he had observed abroad, and devoting himself, above all, to improvements in the rotation of crops, and to the field cultivation of turnips and clover ... So zealous was Townshend's advocacy of turnips as the pivot of agricultural improvement, that he gained the nickname of 'Turnip' Townshend. (1961, p. 174)

Although Ernle admitted that turnips had been introduced into Norfolk in the 50 years before 1730, he maintained that 'even in Townshend's own county, it was not till the close of the century that the practice was at all universally adopted', while outside

Norfolk 'both landlords and farmers still classed turnips with rats as Hanoverian innovations, and refused their assistance with Jacobite indignation' (1961, p. 175). Leaving aside the emotive language, the message was clear: turnips were pioneered in (certain parts of) Norfolk, from whence they spread slowly to other parts of the country to bring about a transformation of agriculture after about 1760.

Unfortunately, Ernle overlooked evidence which suggested that the changes he had dated beyond 1760 had earlier antecedents. Contemporaries were extolling the virtues of turnips and clover well before 1760. In 1682 John Houghton had noted 'pasture lands improved by clover, sainfoin, turnips, coleseed, purslane, and many other good husbandries, so that the food of cattle is increased as fast, if not faster, than the consumption' (Thirsk, 1976, pp. 133−4). Daniel Defoe wrote of East Anglia in the 1720s:

This part of England is also remarkable for being the first where the feeding and fattening of cattle, both sheep as well as black cattle with turnips, was first practised in England, which is made a very great part of the improvement of their lands to this day; and from whence the practice is spread over most of the east and south parts of England, to the great enriching of the farmers, and increase of fat cattle. (Defoe, 1971, pp. 82−3)

William Allen, writing in 1736, noted that it was in the reign of James I that innovations were first introduced on to the arable, and that it was in Charles II's reign that these changes spread most rapidly (Allen, 1736).

In the years after Ernle's book was published similar evidence was gathered in various regions of the country and by the 1960s it was clear that Ernle was wrong about the timing of innovation. A major technological transformation had taken place before 1760; indeed, Eric Kerridge went so far as to argue that 'the agricultural revolution dominated the period between 1560 and 1767 and ... all its main achievements fell before 1720, most of them before 1673, and many of them much earlier still' (1967, p. 328). For Kerridge, Turnip Townshend came at the end and not at the beginning of the Agricultural Revolution, but he also argued that turnips and clover were not the only innovations of importance. This in turn raised questions not merely about when the Agricultural Revolution occurred, but also about the nature of the innovations connected with it.

14

Kerridge identified several developments in the sixteenth and seventeenth centuries which he regarded as critical in explaining the transformation of agriculture. These were the floating of the water meadows (particularly in the south-western counties), 'the substitution of up-and-down husbandry for permanent tillage and permanent grass for shifting cultivation, the introduction of new fallow crops and selected grasses, marsh drainage, manuring, and stock-breeding' (1967, p. 40).

Kerridge was undoubtedly right to point to the early date at which some of these innovations were first employed. A case in point was 'convertible' husbandry, which was an alternative to the traditional rotation pattern of arable followed by fallow. Each field might be cropped for three or four years in succession and then laid down with legumes or artificial grasses (leys). After a few years it was then ploughed again for corn crops. The system was adopted in both open fields and enclosures with the intention of increasing pasture and reducing fallow while still restoring fertility. In the open fields the areas under grass would be grazed by animals such as cattle and horses which could be tethered (and which would not therefore wander on to adjoining strips which were being cropped). Convertible husbandry was particularly appropriate in areas of heavy soils, in places where different crops were grown in the same field, and in parishes where the open fields were grassed over and older sheep walks and rabbit warrens were ploughed up. It certainly predated the eighteenth century; indeed, it probably developed before 1560 and spread through much of Midland England in the following century. However, it was not necessarily a permanent change, and after 1650 some areas reverted to permanent pasture (Broad, 1980).

Why this was so is not clear, but it may have been related to population growth. The population of England grew from approximately 2.8 million in 1540 to 5.2 million by 1650, but no further significant growth occurred until the 1720s. It is possible that changes in agriculture to that date had failed to increase the output of food in line with population growth, which was dampened down accordingly. If this was so, one of the obvious problems arising in connection with Kerridge's dating of the Agricultural Revolution was that the innovations down to 1650 had failed to raise productivity to new and sustained levels. This is not to detract from the importance of his argument that pressure on the resources of agriculture in the shape of land hunger, excess demand for basic foodstuffs, and congestion on (and probably overuse of) grasslands, almost certainly stimulated the reorganization and improvement of practice in this period.

Kerridge may have overstated his case by failing to make a clear distinction between the first knowledge of an innovation and its adoption on a significant scale. Water meadows provide a good example. The floated water meadow was developed on chalkland areas. A dam, and various ducts and drains were constructed so that meadows could be flooded in winter. The chalk and other minerals suspended in the water acted as a fertilizer and the water itself was an insulation against frost. Earlier grazing was then possible. Kerridge correctly dated the introduction of water meadows to the late sixteenth century. However, on the chalk downlands of Wessex, although they became vital to the farming economy in the post-1640 period, before that time they seem to have been neither as widespread nor as efficient as Kerridge believed (Bowie, 1987).

Whatever the potential weaknesses of Kerridge's case he was right to highlight the extent of change before c.1750. Since 1967 his view has been confirmed by the analysis of sixteenth and seventeenth century probate inventories. These were listings of the personal estate of an individual, drawn up shortly after his or her death in conjunction with the granting of probate. They were usually attached to a will or − in the event of intestacy − letters of administration. Agricultural historians have found inventories to be a rich source because they include information about crops, stock and agricultural implements owned by the person who had died. Inventories have revealed the regional diversity of farming practices during the early modern period, and they also provide important evidence about the spread of the new crops. In Norfolk and Suffolk, for example, they have been used to show that there was 'a dramatic upsurge in the adoption of root crops, from a rate of about 5 per cent of farmers from the 1580s through to the 1660s, to 50 per cent of farmers by 1710. The period of innovation for clover was similar, although the proportion of farmers adopting was smaller' (Overton, 1985, p. 211). Similar conclusions have been reached for Hertfordshire where grass substitutes and root crops were listed most frequently in inventories after about 1670 (Glennie, 1988a). Inventory studies have also revealed that clover was grown in the vale of Taunton Deane as early as 1679, lucerne in Cornwall by 1722, and sainfoin in Wales in the 1660s (Thirsk, 1984−5, vol. I, pp. 364, 366, 417, 419).

Historians no longer doubt that innovation, in the form of the growing of turnips and legumes, commenced much earlier than Ernle claimed, nor that the Agricultural Revolution was a rather more complex business than was once believed. The critical period for innovation was probably the century or so after 1650. New crops were introduced from the middle of the seventeenth century (Thirsk, 1967) and gradually encompassed all areas, although even in East Anglia the use of fodder crops still had a long way to go in 1760. Turnips represented only about 8 per cent of the cropped acreage in the 1720s, and clover just 3.5 per cent. By the 1850s these proportions had grown to 18 and 20 per cent respectively (Overton, 1983). Further afield, in areas such as Cumbria, the new husbandry had only a tenuous hold by 1760, while in the Midlands inadequate drainage hampered the adoption of the new husbandry and in many places open-field, mixed farming was replaced by permanent pasture.

Turnips and clover were not the only new crops. Potatoes, tobacco and hops provided important new cash crops; market gardening was a significant innovation from the 1590s and played an important role in feeding the population by the mid-eighteenth century; fruit farming spread through Worcestershire, Hertfordshire and Kent. Other changes included the substitution of barley for rye. However, terrain and tradition meant that improved farming spread erratically, producing a patchwork effect with progressive and backward practice taking place in neighbouring regions if not adjoining farms. In north Lincolnshire, Draynor and Burrell Massingberd were using clover and turnips on their estate at the end of the seventeenth century, but generally this was a backward area until the 1820s and 1830s (Holderness, 1972). Some of the best cultivated farms in the country were to be found on the east Yorkshire wolds by the nineteenth century, but the old ways persisted in the adjoining lowlands where, as late as 1850, rotations characteristic of open-field farming were still practised in some enclosed parishes (Harris, 1961). As Joan Thirsk has noted of the period before 1750, 'in many regions of England the problems of finding a satisfactory farming scheme within the traditional framework were insoluble' (Thirsk, 1984–5, part II, p. 587).

As a result of these gradual changes, the most critical development of the period 1760–1830 was the spread of the techniques and their widespread adoption. It took time to persuade farmers to adopt new practices, and, similarly, it was many years before they were fully effective. Legumes, for example, produced a gradual build-up in total soil nitrogen, and it would have taken several decades from the first introduction of clover in place of fallow for a new ceiling of fertility to be reached (Chorley, 1981). This is not to imply that the post-1760 period was entirely devoid of innovation. Particular problems continued to call for solutions. A good example concerns the common turnip, which had a tendency to freeze in particularly cold winters, and then to rot during the subsequent thaw. In hard winters it was unreliable as a resource for feeding an expanded national sheep flock, and the adverse weather of 1794—5 and 1798—9 encouraged the adoption of the hardier, less watery and therefore more frost-resistant swede. Various Herefordshire farmers were soon experimenting with the swede, and by 1805 it was reported to have made 'very considerable progress' in the county (Jones, 1974, p. 51).

The war years 1793—1815 also provided encouragement to farmers who were reluctant to update their practices. In addition to expanding — albeit temporarily — the cultivated acreage, farmers looked to raise output through adopting innovations they may previously have neglected. In south Lincolnshire turnips and artificial grasses were introduced on a considerable scale, and the long-term effect was one of very substantial improvement in local husbandry by the mid-nineteenth century (Grigg, 1966).

These were changes within a relatively established farming system, and further technological advances were delayed until the 1830s. However, nineteenth century changes were of a different nature from their predecessors. In the past, output had been raised by introducing new crops; after c.1815 the progress of agriculture depended on increasing the fertility of the soil, and considerable effort went into improvements which could be achieved by exploiting existing practices to the full.

Traditionally, farms were self-sufficient in the sense that farmers produced their own hay, straw and grass to feed their animals, and in turn used their waste products as manure. This was not an invariable rule. From the sixteenth century, lime, in limestone areas, was burnt in a kiln and then spread on the land to sweeten acidic soil and raise fertility. Liming was adopted by the more progressive Devonshire farmers between 1550 and 1650, and by 1750 it was widespread in the county (Havinden, 1974). Elsewhere, marl was spread on the ground to give body and retentiveness to light soils and to make heavy soils workable, although overuse could produce diminishing returns. Pulverized slag from ironworks, soot, coal ashes, waste brine from saltpans, and even rags, were among the different materials used by enterprising farmers to try to improve the fertility of their land. Farmers near towns purchased 'urban night soil'; Suffolk farmers in the later eighteenth century 'get their muck from London and a sort of chalk they mix with it from Kent: they arrange for the muck to come from London on the barges that take there the local produce and that otherwise might return empty'. Not surprisingly perhaps their lands 'never lie fallow but are cropped in the four-year rotation' (Scarfe, 1988, p. 126).

From about 1830 these isolated examples of enterprise became more common as farming awoke from nearly two decades of marking time after the Napoleonic wars. The 1830s and 1840s witnessed the widespread adoption of the Norfolk four-course rotation as part of a significant intensification of cultivation (Brown and Beecham, 1989). Farmers gradually ceased to regard the farm as self-sufficient in the supply of manure and feedstuffs, and began instead to purchase new manures, particularly bones and rape cakes, and feeding stuffs such as maize and cotton-seed cake. In a sense, the operations of the farmer became more like those of the factory owner. Farming changed from being an occupation primarily concerned with extraction from the soil into one involving the purchase of raw materials which were processed to produce a saleable product. This took place on a scale sufficiently large to constitute what F. M. L. Thompson has termed a second Agricultural Revolution (Thompson, 1968).

The term 'High Farming', which is usually associated with this second Agricultural Revolution, was widely used from about 1850 to describe almost any kind of progressive, capital-intensive farming. It was synonymous with high levels of output which were achieved by the judicious application of the new knowledge available to farmers. High Farming was particularly associated with the 1840s, a decade notable for a number of critical developments including the import of guano (excrement of sea-fowl from Peru) and nitrate from Chile for use as manure, the introduction of superphosphate and of the first 'artificial' or chemical fertilizer on a commercial scale, and the founding of, respectively, the Rothamsted Experimental Station in 1843, the Royal College of Veterinary Surgeons in 1844, and the Royal Agricultural College at Cirencester in 1845. The Royal Agricultural Society started its annual journal in 1840. Farming, in other words, was becoming more scientific. On the other hand, vast gaps remained to be filled and farmers preferred tried and tested methods to new-fangled ideas described in books and journals.

If the commercialization of the farmer was one important change in the middle decades of the nineteenth century, a second was the attention paid to the claylands. In the post-Napoleonic war years the clayland wheat farmer was badly hit by falling grain prices. Clayland farmers depended on the wheat crop to pay the rent, but after 1815 − partly as a result of practices designed to push up output during the war − both their grassland and arable were exhausted. The normal pattern of farming was for farmyard manure to be spread on corn land. The meadows were manured only during periods of the year when stock were grazed on them. To increase the number of stock, farmers needed to grow fodder crops, but the moisture retentiveness of the clay soil made this almost impossible. Turnips, in particular, could not be grown either for feeding the animals through the winter or for putting stock in the field to feed.

To break out of this cycle the clays needed to be drained. Nathaniel Kent, a leading land surveyor in the second half of the eighteenth century, wrote in 1775 that 'draining is the first improvement that wet lands can receive' (1775, p. 17). Traditionally, farmers ran off water by ridge and furrow ploughing. Land was ploughed in such a way as to raise ridges at regularly spaced intervals, producing a corrugated effect in the landscape, and the rainwater was carried away through furrows. One stage on from this was for farmers to dig trenches which they filled with stones and other matter. Neither system was as effective as tile drainage, but this was rather more expensive. The normal pattern was for landlords to supply the tiles and the farmer to lay them. Something more effective was required and it arrived with the invention in the 1840s of inexpensive, machine-made cylindrical pipes, together with a major boost to the farming community in 1846 in the form of cheap loans. Seen by contemporaries as a sop to the landed interest for the repeal of the Corn Laws — which were thought to have maintained grain prices at an artificially high level since 1815 — the drainage schemes played an important role in farming, particularly on the claylands. Between 1846 and the agricultural slump of the 1870s around £24 million was invested in drainage and related projects, of which £4 million came from the government, £8 million from private companies and the rest from the owners' own resources (Clapham, 1939, vol. II, pp. 271–2).

The impact of drainage on clayland farming was considerable. Mixed arable was converted to intensive grassland husbandry. By feeding cows with grains and oilcake, meadowland was released for summer grazing, and expanded herds of dairy and beef cattle could be carried on summer pasture. In turn this increased the available dung, giving higher yields of hay and corn for feed, and permitting a reduction in the meadow and arable acreage. Fewer hay-consuming plough horses were needed, so that more land was laid down for summer pastures. Improvement was directed towards expanding dairying and grazing on land where grass was a natural crop. The result was an increase in the yield of the grasslands. In

20

addition, vetches and mangolds were grown, permitting an extension of grazing land at the expense of meadow, and making it possible for farmers to carry larger dairy herds and more fattening stock on summer pastures (Sturgess, 1966, 1967).

Too much should not be made of this for the profitability of farming. Pipe drainage was an important break-through for the clayland farmers, although historians have had doubts as to how far landlords received a return on their investment. The real increase in the physical output of livestock and livestock products after 1850 has been questioned; indeed, it has been suggested that the changes in clayland farming practice produced little more than unsatisfactory adaptations to market trends in favour of livestock production (Collins and Jones, 1967). This is probably going too far in the opposite direction, although detailed studies of drainage schemes have tended to confirm their lack of profitability (Phillips, 1969, 1975). Above all, these developments produced a farming balance which was profitable as the market expanded in the mid-nineteenth century, but which proved vulnerable to the post-1873 collapse in grain prices.

High farming and drainage schemes were expensive, and a feature of improved farming was that it required a greater input of capital from the farmer. The tenant's capital requirement for the new husbandry has been estimated to have been 40 per cent greater than under the regime of two crops and a fallow, while contemporaries believed that the investment level on a successful farm doubled during the French wars (Jones, 1981). These costs rose again during the second Agricultural Revolution. Drainage was obviously expensive, but considerable sums also went into improved farm buildings and accommodation, and these costs fell almost exclusively (particularly in the case of buildings) on the landlord (Holderness, 1988). In Oxfordshire, investment in all forms of long-term capital formation in agriculture (except enclosure) were at their most intense after 1850, and as a result the major changes in the nature, rate and scale of agricultural development occurred in the period 1840–60 (Walton, 1976). This was just one county, but the general picture of these decades was of more efficient, capital-intensive, and highly productive farming (chapter 5).

Livestock Farming

In the seventeenth century most areas of the country practised some form of livestock farming. Upland areas were renowned for their concentration on animals at the expense of arable husbandry, but elsewhere sheep and cattle played an important role in destroying weeds and providing manure for the arable. Cheese, manufactured in the farm dairy and sold at the weekly market, provided the farmer with much needed cash flow. However, despite their undoubted importance within the farm economy, the animals were seen as playing a secondary role compared to the prime importance of the corn crop. It was not until the middle decades of the nineteenth century that farmers began to appreciate the extent to which livestock production could be profitable in itself rather than being merely an adjunct to cereal cropping (Jones, 1968). This conservative attitude towards the farm animals partly accounts for the slow pace at which selective breeding was adopted, although this was also a result of the relative backwardness of the biological sciences. Progress depended on farmers experimenting with breeding practices which often ran ahead of genetic and veterinary science. The attitude was also reflected in the lack of attention paid to grasslands. Arthur Young, for example, thought that permanent pasture farming was a poor and regressive system of management. Not every farmer agreed. As early as the 1680s, on the Verney estate in north Buckinghamshire, fertilizers spread on grasslands included dung, urban night soil, river mud, marl, lime and potash (Broad, 1980). In general, however, it was well into the nineteenth century before much attention was paid to the quality of grass.

For most livestock farmers the emphasis was on fattening rather than on breeding. Traditionally, cattle and sheep were brought 'on the hoof' from Wales and Scotland, and then fattened in the Midlands, East Anglia and the Home Counties in preparation for the London market. The trade in Welsh cattle dates back to the Middle Ages, and by the seventeenth century the Scottish trade was also substantial. Over 150,000 Scottish cattle and a similar number of sheep were exported to England in the nine years 1681−6 and 1689−91, and a further 178,000 cattle crossed the border between 1696 and 1703. Even so, supply may not have kept pace with demand, hence the import of Irish cattle and sheep between 1679 and 1681 when the ban on such trade was temporarily lifted, and the growth of the Scottish trade after the Act of Union in 1707 (Woodward, 1977). After 1707 no customs duties were demanded, and detailed figures do not survive, but by the middle of the eighteenth century 80,000 cattle and 150,000 sheep were being moved south. The small, black Scottish cattle were driven from their breeding grounds in north and north-west Scotland, fattened in the north of England and eventually sold on the London market.

Animal by-products were used extensively. The English wool trade had flourished in the Middle Ages, and woollen cloth was a major product of the West Riding throughout the Industrial Revolution. Cheese was manufactured on a substantial scale in Gloucestershire, Wiltshire and Cheshire, and much of the produce went to London. Among other farm animals, turkeys and geese were driven to London from Norfolk and Suffolk. The extent of the trade is unquantifiable, but its significance was reported at length by Daniel Defoe in the 1720s.

Improvements to livestock required selective breeding, the creation of new breeds, or the improvement of native varieties by crossing with newer varieties. Crossing of animals to improve native breeds had taken place for centuries as farmers sought to improve the weight and quality of their beasts in order to fetch higher prices in the market place. Attempts to increase the size of sheep and the length of the wool staple dated back to the days of the medieval wool trade, but scientific selective breeding began only in the early eighteenth century. By the 1720s Daniel Defoe could write that:

> The sheep bred in [Leicestershire] and Lincolnshire are, without comparison, the largest, and bear not only the greatest weight of flesh on their bones but also the greatest fleeces of wool on their backs of any sheep of England; nor is the fineness of the wool abated for the quantity; but as 'tis the longest staple so 'tis the finest wool in the whole island ... These are the funds of sheep which furnish the City of London with their large mutton in so incredible a quantity. (Defoe, 1971, pp. 408–9)

The most famous name associated with sheep breeding was Robert Bakewell, one of Ernle's heroic figures.

In 1790 the agriculturalist William Marshall wrote that 'to the ability and perseverance of Mr Bakewell the Leicestershire breed of sheep owes the present high state of improvement' (1790, vol. I, p. 382). The principle of selective breeding was known before Bakewell began his livestock experiments in 1745. His work was not unique, but his contemporary eminence arose from the fact that he was solely a specialist breeder and not a grazier-cum-breeder. Bakewell probably borrowed from racehorse breeders the

23

pedigree concept and its use as a basis for breeding value judgement. He selected more rigorously than some of his contemporaries, and he was more sophisticated than they were in the nature of the characters he selected from. Earlier breeders had not had the foresight, patience and resources to breed exclusively from the finest animals. Bakewell's fame was established by the 'New Leicester' breed of sheep, which fattened rapidly and had a high proportion of saleable flesh to bone (Russell, 1986). It fulfilled his main economic objective which was to create a profitable meat animal which would grow quickly, since a higher stock turnover meant a more efficient use of grazing resources. However, his achievements should not be exaggerated. The real difference between him and other breeders was in the general appearance and superior growth rate of his sheep, and the new Leicester was not an unqualified success; the breed was unsuitable for exposed environments such as Romney Marsh, and its flesh was described as inappropriate 'for genteel tables', or more explicity as 'coal-heavers' mutton'.

Bakewell's primary success was with sheep but he also played an important role in cattle breeding. By applying the same principles to his cattle as had given him success with the new Leicesters, Bakewell produced New Longhorns. Here, however, he was following where others had already trodden, and although his new breed put on considerable quantities of fat, it was not a good milker. As with his New Leicester sheep his New Longhorn cattle did not survive in the way he had bred them, but his example was followed by others. The most famous of these followers was George Culley, the celebrated Northumberland farmer, who bred from Bakewell's stock (Macdonald, 1979b).

Elsewhere, in about 1780, John Ellman of Glynde in Sussex refashioned the shortwool breeds in a manner similar to Bakewell's efforts with longwool sheep. Also in the 1780s, the Colling brothers, who farmed near Darlington, were responsible for the successful inbreeding of cattle, when Bakewell's longhorn gave way to their improved shorthorns. However, these pioneers often worked with only a handful of followers so that it was not until well into the nineteenth century that the importance of quality in livestock spread from the progressive few to the general run of farmers (Trow-Smith, 1957, 1959).

Intensive mixed husbandry brought a substantial increase in sheep numbers, since folding or ranging sheep on turnips or clover leys brought advantages in dung and secondary commercial commodities. Improvements in feed, with the introduction of forage crops, increased both the number and weight of sheep in mixed farming systems. New breeds did not automatically bring carcass weight increases but they did cut the age of slaughter which increased the supply of meat and in turn reduced the price. The new breeds of sheep were ready for the butcher in two rather than four years. As animals were fed for less time before market, and the amount of saleable flesh from each animal increased, the price of meat fell. Contemporaries believed that this produced an increase in output. On the other hand the number of animals may have fallen slightly during the French wars as pasture land was ploughed to extend the cropped acreage (Mingay, 1989, pp. 313–51).

The major improvements in breeding are usually associated with High Farming. With rising living standards and the building of the railway network, urban demand for meat and dairy products increased. The expansion of the market for liquid milk took place so rapidly that supplies of cheese and butter had to be imported from France, Holland and America. English farmers began to move their interests away from the old corn staple towards meat, milk and vegetables. In other words, High Farming was characterized by a shift of demand away from bread towards meat and dairy products, partly because the higher prices of meat products produced market conditions conducive to a decisive swing from wheat growing towards livestock production. However, it was not a wholesale switchover; rather the output of animals simply expanded within the context of mixed farming and the extent to which the animals themselves were improved remains open to question (Walton, 1986).

Labour and Machinery

A central characteristic of the Industrial Revolution was the re-placement of labour by machinery. The same process in agriculture partly depended on progress in industry, particularly the substi-tution of cast-iron standardized implements for wooden ones, made by the village blacksmith and carpenter. This in turn depended on developments in iron founding. However, the introduction of new tools and machines was remarkably slow in agriculture; indeed, it is no exaggeration to say that labour-saving machinery was widely adopted only in the twentieth century. While this was partly because technological progress occurred slowly and farmers were sometimes unable to afford to buy machines, it was predominantly a consequence of the labour market.

As a general rule farmers were unwilling to invest in machinery while labour was cheap and plentiful. It is significant that machinery made quite rapid progress during the French wars, when farmers were confronted by labour shortages. The problems were such that one company of Welsh militia, most of who were rural labourers by trade, were induced to go into the fields around Whitehaven where they made use of the scythes 'they providently brought with them' (Jones, 1974, pp. 213−14). Labour shortages brought high wages and encouraged the adoption of alternative hand tools, but after 1815 this process slowed down as agricultural wages fell. Labour productivity (and wage rates) remained lower than in industry because of the relative immobility of labour. It was this which convinced many farmers that there was little point in adopting machines, even those of proven ability and offering a clear cost advantage because the social costs might outweigh the economic benefits. Machines designed to speed up the process of harvesting were sometimes used only for wheat, leaving barley and oats to be cut or beaten out of the land. However, this is not to minimize the extent to which farmers capitalized on the labour surplus to pay low wages: 'to some extent, at least, the productive achievements of [1750−1850] were secured at the expense of the hardship and deprivation of the more than 900,000 workers who laboured on the farms of England and Wales for meagre rewards' (Mingay, 1989, p. 961).

Farmers who did try to introduce machinery often encountered hostility. The Captain Swing riots of 1830 focused particularly on the hated threshing machine (Hobsbawm and Rude, 1969), but less well known is the constant drip of machine-breaking − particularly of harvest machinery − which went on through the following decades. Partly as a result, landlords were concerned that all available labour should be employed. The alternative, particularly during the winter months, was to support those out of work with poor relief. Any shortfall in labour during the harvest period could be accommodated by hiring casual labourers, many of them from Ireland, who roamed the countryside in search of temporary work.

This responsibility of the community to provide subsistence for as large a number of people as possible ensured that technological unemployment was not a major social problem in English agriculture during the nineteenth century. As E. J. T. Collins has noted:

as late as 1870 much the greater part of the British corn harvest was still cut by hand tools, even though other farm operations such as threshing and livestock feed preparation, which mostly occupied the off-peak periods of the farming year, were already highly mechanized. (Collins, 1969, p. 455; 1987)

It was considerations relating to the labour market which brought about this apparently contradictory situation where machinery was least used during some of the most labour-intensive parts of the farming calendar.

The reluctance of the farming community to adopt machinery inevitably dampened innovation, but plenty of machines came on the market over time even if they were not widely adopted. The iron plough was the first major improvement on the traditional wooden implement. By the end of the eighteenth century the Rotherham plough was coming into widespread use, particularly in the north and east of the country. So-called because it was first used on a farm near Rotherham in south Yorkshire, the plough was introduced into England from Holland around 1730. It had a much lighter framework, and an iron-plated mouldboard (the part which turned the soil). Self-sharpening, hardened cast-iron plough-shares were introduced by Robert Ransome in the 1780s and patented in 1803. Seed drills and horse-drawn hoes had been invented in the late seventeenth century, but they were adopted only very slowly before improvements were made to Jethro Tull's early designs in the 1780s. Even then it was well into the nineteenth century before seed drills were widely used; indeed it was only during the 1830s and 1840s that they were adopted nationwide (Smith, 1984). Other innovations dating from the 1780s included machines for threshing, chaff cutting, root slicing and crushing. Few of these were widely adopted, and most used hand or horse power rather than steam power.

Technological experiments were usually concentrated on the most labour-intensive part of the agricultural cycle, hay-time and harvest. Many designs for mechanical reapers were patented in the eighteenth century but none were introduced successfully. It was the 1820s before the first successful reapers came into widespread use, but hay and grain continued to be cut by hand. By 1850 the more efficient scythe had largely replaced the sickle. Threshing, the process of separating the grain from the straw, traditionally took place by hand using a flail. This was another time-consuming task which occupied many days through the winter and spring. The first widely used threshing machine was introduced by a Scotsman, Andrew Meikle, in 1786. It could be driven by steam, water, horse, or hand, and together with the slightly later winnowing machine (to blow the chaff from the grain) it had a profound impact on this particular farmyard activity. Five work-days were needed to thresh the produce of one acre of wheat using a flail; a steam-powered thresher could do the same task in less than one work-day. From the 1780s men began to travel the countryside with mechanical threshers, and to rent their skill and machine by the day, making it possible for even the smallest farmer to afford the service (Collins, 1969; 1976).

High Farming brought further technological advances. The Royal Agricultural Society, founded in the 1830s, had among its principles the 'improvement of agricultural implements', and during the 1850s new reaping machines came on the market, together with mechanical binders. The first hay mowers dated from about 1860 as did the first steam engines used in farm work. The 1850s and 1860s were decades when Lincoln, Grantham and Gainsborough developed extensive engineering interests in powered mechanical implements for agriculture, making steam engines, drainage pumps and portable threshing machines. Ransome introduced the first successful multi-furrow plough.

Despite these steps forward the impact of machinery in agriculture should not be exaggerated. Part of the problem was the traditional conservatism of the farming community, and it is significant that many of the major agricultural engineering companies which flourished after 1850 depended for their success on exports. Until wages began to rise in the middle decades of the nineteenth century farmers were reluctant to pay serious attention to technological innovation: the Agricultural Revolution was not a result of mechanical ingenuity.

Why did Innovation Occur?

Ernle's view was that innovation began with the landlord, hence his insistence on the critical role of men like Coke at Holkham, the Duke of Bedford at Woburn, and Lord Egremont at Petworth. Through the example of their home farms, and the encouragement offered by their sheep shearings and other farming festivals, they set an example which permeated down the farming ladder to the general benefit of the whole community. On the whole, however, Ernle may have mistaken paternalist endeavour for real achievement. As patrons of agricultural shows and societies, and as promoters of Parliamentary legislation on behalf of the agricultural interest they played an important role. In addition, a handful of leading aristocrats were major figures in agricultural improvement. They included the larger landlords of north Nottinghamshire, all of whom had progressive home farms from the later eighteenth century. James Caird described the Duke of Portland's 400 acres of water meadows in Sherwood Forest as 'the pride of Nottinghamshire, unrivalled as a work of art in irrigation, and in its cost worthy of the liberality of a wise and patriotic nobleman' (1852, p. 206).

On the other hand plenty of landowners took little interest in their resources. In the Lindsey division of Lincolnshire most of the larger owners had no conscious policy of improvement, while Caird noted of Oxfordshire in the 1850s that, 'as a general rule the landlords of this county interest themselves very little in agriculture. Few of them are practically acquainted with, or engaged in farming' (1852, p. 27). Absenteeism was an obvious reason for poor farming, but it was not necessarily the case that estates went to rack and ruin when the owner was away. Indeed, an efficient steward might turn out to be rather more effective in promoting the cause of good farming than an untrained and perhaps even unwilling resident owner (Beckett, 1983b).

The true credit for agricultural innovation ought perhaps to rest with the lesser landowners, with estate stewards and with tenant farmers. The gentry pioneers of improved practices have long been recognized (Thirsk, 1974). In Herefordshire they 'formed the spearhead of innovators'; as John Beale wrote of them in 1657, 'I observe the wisest and best of our gentry to be very careful in setting forward such kind of husbandry, as agrees with the nature of the soil where he inhabiteth' (Jones, 1974, p. 48). Stewards have usually received a good press for their work in encouraging improvement, despite their almost universal reputation among contemporary landowners for dishonesty. In the course of the eighteenth and nineteenth centuries the position of steward became formalized as 'agent' and professionalized, particularly from the opening of the Royal Agricultural College at Cirencester in 1845.

Tenant farmers, the men who actually undertook the work of innovation, played a vital role. Coke's reputation depended as much on his tenants as on his home farm (Parker, 1975). Generally it was the more substantial farmers who led the way, and their lesser brethren who followed. George Culley of Glendale in Northumberland was an improving farmer whose example encouraged imitation, although many of those who followed him seem to have been more interested in his ways of making money than his agricultural techniques (Macdonald, 1979b). In East Anglia the adoption of both turnips and clover moved down the social hierarchy through time, with the smaller farmers marching in the rear (Overton, 1985). In Oxfordshire over the period 1750−1850 substantial tenant farmers were as important as large landowners in promoting pedigree shorthorn breeding (Walton, 1976); while the success of the Hereford breed of cattle was attributed to 'dirty-boot tenant farmers' (Jones, 1981, p. 78).

Nor is this picture surprising. Farmers were naturally conservative. Despite a voluminous farming literature, they preferred to rely on the firm recommendation of other farmers rather than the virtues of new techniques as extolled in newspapers and periodicals (Macdonald, 1979a; Goddard, 1989). Francois de la Rochefoucauld commented of English farmers in 1784 that they 'gather all kinds of information from their neighbours and from those who have better results than they have themselves. The clubs where they often go provide the opportunity of instructive conversations' (Scarfe, 1988, p. 151). By these means small farmers were capable of making a significant contribution to the progress of agriculture even if it was in the wake of examples set by gentry and better-off farmers.

The critical question, however, is not who promoted innovation but how was it that farmers came to work their holdings by more efficient and productive means? Was it simply a result of the operation of the price mechanism, or did it also have something to do with the relationship which existed between landlord and tenant in the English countryside?

Throughout Europe the period 1650−1750 or so was characterized by stagnant and even falling grain prices, which in turn reflected trends in population. Only in two countries, Holland and England, does the response to this movement appear to have been progressive farming. The significance of this discontinuity cannot be underestimated. Coleman has argued that England's economic performance suggests that improvements in agricultural productivity in the 1650−1750 period were such as to represent a real divergence from the rest of Europe (1977, p. 199). But why, in apparently adverse conditions, did English farmers differ so markedly from their Continental neighbours?

English farmers compensated for falling grain prices by switching their attention to animal products, but this brought the rather perverse result of increasing grain output. The introduction of fodder crops enabled farmers to keep more animals. More animals meant more manure, and more manure meant more fertile soil. Fodder crops cannot have been introduced with the intention of increasing grain yields. Although the Norfolk four-course raised yields, by contrast with a three-course system less grain is actually grown: 'unless farmers increased the total area cultivated they would have needed great faith that an increase in yields would not be offset by a fall in the area of grain' (Overton, 1983, p. 3).

In East Anglia turnips may even have been introduced to provide an alternative and higher yielding source of fodder because the grain acreage had been extended to grow more barley at the expense of permanent pasture. Turnips were probably planted in the summer, when the hay harvest could be judged, as an insurance winter fodder crop. The late seventeenth century is sometimes referred to as the Little Ice Age, and at least one churchwarden noted that the want of hay in 1681 was underwritten by growing turnips. It may have been, therefore, that fodder crops were introduced in order to keep more animals − as a result of prices − and that the advantages in terms of grain became apparent only subsequently. It is also somewhat paradoxical to suggest that farmers were squeezed by rising costs, among them labour costs, but were introducing new fodder crops which were labour intensive. Consequently it seems unlikely that fodder crops were introduced with the intention of cutting costs by pushing up yields, even if this turned out to be the result.

An increase in grain output could even have been disastrous for the farming community but for two related developments. One was government support. The 1670 Corn Law introduced a sliding scale of duties on imports of wheat, determined by the selling price at home. The Corn Bounty Act of 1688 subsidized corn exports to keep prices above 48 shillings a quarter. They were suspended in years of dearth (1698, 1707, 1740, 1757, 1767) when exports were prohibited. As a result, between 1697 and 1701 43 million quarters of wheat were exported, of which $c.24$ million (55 per cent) went abroad in the period 1732−66, and nearly 400,000 quarters annually in the 1740s (Ormrod, 1985; Ashton, 1972; Chartres, 1984−5). Coupled with government efforts to promote the use of grain for malt, in beer and in spirits, the effect was to protect the farmers from the worst effects of their own overproductivity, since whatever the cause of the increase in output it must further have depressed prices (see figure 1, p. 65).

The second related development arose from the success of grain farmers on the lighter soils. Their prosperity highlighted the plight of those struggling to make ends meet on the heavy Midland clays. Here also the answer to the farmer's prayer seemed to be in concentrating on meat and dairy products, but since fodder crops

31

were impractical landowners instead turned down their land to grass. Consequently, a further effect of agricultural conditions in the period 1650–1750 was a steady shift in arable farming away from the heaviest and worst-drained clays (the ancient cornlands of England) towards the lighter, sandy and loam soils, many of which in the past had been heaths or commons. The knock-on effect was of an increase in cereal output per acre – especially as it was these areas which were the most appropriate for alternate husbandry – and an improvement in the quality of grazing for livestock because grass grows better on moisture-retaining clays than on dry heaths.

These changes could not have occurred without a close working relationship between landlord and tenant farmer. English landowners were burdened with the fixed capital costs of running their estates, leaving the tenant to use all his capital in running the farm. As a result, landlords had to find the cash for improvements such as enclosure, drainage, replacement of farm buildings, and general maintenance between tenancies. Except in areas where older tenancy arrangements still persisted, they were also in a position to move land into the hands of their better tenants and to write progressive husbandry clauses into leases.

Eighteenth-century commentators generally favoured long leases on the grounds that they offered security to the tenant and scope for making changes. The ideal ones were those at Holkham. However, the fluctuations of the war years 1793–1815 turned many landlords against long leases because they were unable to raise rents in a period of inflation. Moreover, where landlords dispensed with leases tenants were not noticeably less efficient, and, despite claims that they were the only guarantee of agricultural improvement, long leases began to disappear. In the course of the nineteenth century they were widely replaced by annual tenancies with agreed tenant right.

In any case there was little evidence that long leases had played a vital role in promoting agricultural improvement, and annual tenancies were advantageous to owners and farmers alike. Tenant right, the sum of money paid to an outgoing tenant in compensation for his unexhausted improvements, was a sufficient guarantee for most enterprising tenants. Farmers enjoyed effective security of tenure, and were able to introduce new fertilizers and to undertake other experiments knowing they would be recompensed for their efforts (Perkins, 1975). Possibly the most valuable role of the lease was in persuading farmers to undertake the risk of cultivating newly-enclosed waste by offering low rent for the early years of the tenancy.

Similarly, few landlords were willing to accept the advice of agricultural writers that they should stretch their farmers by setting the highest possible rents. Even Coke let his farms at a moderate rent, while the agent of the Marquess of Stafford's estate wrote in 1830 that:

> Lord Stafford's rents ... have always been fixed at rather under the general average of the district ... I mean that the tenants should feel that they hold their lands on rather easier terms than their neighbours. It is fit and proper that those who hold of a great man should do so. (Richards, 1973, p. 29)

Overall landlords sought to help and encourage their tenants without resorting to coercion, and these amicable arrangements enabled progressive agriculture to flourish.

Finally, what was the part played in innovation by extraneous factors beyond both the farmers' and the landlords' control? The most obvious of these was the question of marketing. The century or so prior to 1750 was a period of integration and efficiency in marketing which had the effect of accelerating the natural wastage of markets and fairs. This in turn was a consequence of transport changes. Turnpike road improvements widened the market for goods: 'Improved means of road transport were ... a dynamic force in the development of agricultural marketing in the century before 1750' (Chartres, 1984–5, p. 466). Coastal shipping enabled regions such as East Anglia to communicate with markets as far apart as London and Tyneside, and aided the trade in Cheshire cheese to the capital. Although farmers looked primarily to their local market towns in order to buy and sell, the pull of the London market was felt almost everywhere by the 1720s. As urbanization proceeded ever greater amounts of agricultural produce had to be moved distances. Canals played a significant role in the transport of grain, while in the nineteenth century the railway opened up many new opportunities. It was no longer necessary to drive animals 'on the hoof' to market, and thereby to see the product grow thinner on the way, while the movement of fresh milk became possible on a scale not previously contemplated. The specialization of farming which had begun in the seventeenth century proceeded during the eighteenth and nineteenth centuries hand in hand with improved communications.

3 Enclosure

Enclosure, the taking in of new land for cultivation purposes and the enclosure of communally administered landholdings often held in large unhedged fields, loomed large in traditional explanations of the Agricultural Revolution. It was an issue of considerable political importance in the sixteenth century when successive Parliaments passed legislation against enclosing and engrossing. By the eighteenth century attitudes had changed, and historians have attached considerable importance to the 1760–1815 period when enclosure with Parliamentary approval took place on a massive scale. The Parliamentary enclosure movement, so-called, was seen as a consequence of the efforts of landlords and tenant farmers alike to increase the efficiency and output of agriculture. However, this was achieved only at the cost of reducing employment on the land, and many individuals were forced into the towns in search of work. Thus enclosure was good for agriculture, but it had harmful social side-effects. As our knowledge of the Parliamentary movement has increased, so has our awareness of the extent of non-Parliamentary enclosure before 1760. Similarly, the view that enclosure automatically raised productivity has been questioned, while its harmful social effects have been shown to be exaggerated. In this chapter we shall ask what was enclosure, how did it affect the landscape and farming practice, and did it have harmful social consequences?

New land enclosed for cultivation could be of two types. First there was land, sometimes of good quality, which had not previously been cultivated. Considerable effort went into draining the Lincolnshire fenlands in the early seventeenth century and again in the nineteenth century. The area around Spalding was transformed by the introduction of steam pumps in the 1830s and 1840s, and grain production grew so rapidly that the area became a net exporter within a couple of generations (Beastall, 1978). Second, moor, waste or forest land could be cleared and cultivated.

This was poorer quality land and yields would not have been as high as those achieved elsewhere, but it had the effect of pushing up overall output. This type of enclosure was particularly character-istic of the Napoleonic war years.

Enclosure could take place in a number of different ways. Before the eighteenth century it was generally 'by agreement', although what this entailed in practice is not always clear. It really meant enclosure by a private arrangement of the owners of land, and where only one or two individuals were involved it probably did mean just this. Many so-called agreements must have been disputed, and in some cases they were contested in the courts. As a result, agreements were often enrolled in the courts of Chancery and the Exchequer in order to establish their legal authority. Enclosure by Act of Parliament was more expensive but it conveyed legal sanction to the agreement. A petition was presented to Parliament which, if it was accepted, formed the basis of a bill and then an Act. Under the terms of the Act commissioners were appointed to re-allot the land, and their decisions were incorporated in an enclosure award. This might not be completed until several years later. Fences were then erected, hedges planted, new farm houses built, and usually the whole farming structure was altered, particularly when an open-field village was enclosed (Tate, 1967).

Enclosure History 1500–1914

Recent estimates by J. R. Wordie suggest that approximately 45 per cent of the land area of England was enclosed in 1500 and that the proportion rose to 47 per cent by 1600, to 71 per cent by 1700 and to more than 95 per cent by 1914 (Wordie, 1983). These figures are questionable on various grounds. Why, for example, does so little enclosure appear to have taken place in the teeth of political opposition, and when the incentive was con-siderable, during the sixteenth century, and so much in the seventeenth century with hardly a murmur of publicly recorded opposition? What they confirm, however, is that piecemeal en-closure over many centuries had, by the eighteenth century, greatly reduced the acreage of land still farmed in open fields, and that large areas of the country were already enclosed by the time Parliamentary enclosure began in earnest around 1760 (Butlin, 1979).

On the Midland clays, the homelands of open-field farming, enclosure had probably been proceeding piecemeal for a century or more before 1760. It was these areas which were most affected by static and falling grain prices after c.1650, and where landowners sought to offset their losses by turning down arable to grass. In Nottinghamshire, the villagers of Cotgrave enclosed one of their four open fields in 1717 to turn the land over to grass; while in Laxton, still celebrated today for its open fields, between 1727 and 1732 large areas of woodland were cleared and the meadows enclosed to create more grazing land. In north Buckinghamshire enclosure was widespread even if much of it was not recorded in any official statistics. Of 138 parishes in a sample area, 52 were entirely enclosed without an Act of Parliament, and evidence has been found of pre-Parliamentary enclosure in at least half of the other 86 (Reed, Michael, 1984). The period 1660–1760 saw a steady transformation of Leicestershire from arable to pastoral farming. Between 1660 and 1710, 41 villages are known to have been enclosed and a further 32 during the years 1710–60. More than half the county was already enclosed in 1730 (Thirsk, 1954).

Although the first enclosure by Act of Parliament was in 1604, it was the period 1750–1830 which is most closely associated with Parliamentary enclosure. In these decades more than 4,000 Acts were passed, permitting the enclosure of approximately 6.8 million acres (i.e. 21 per cent of the total area of England), and this may be an underestimate (Chapman, 1987). The chronological co-incidence of Parliamentary enclosure with the Agricultural Revolution suggests a causal relationship; thus McCloskey has written that 'the eighteenth century then, in the second half of which Parliament added broad powers of compulsion to the tools available for dismantling the open-field system, is the pre-eminent century of English enclosure' (McCloskey, 1975). This is clearly an exaggeration. Ernle, and E. M. Leonard before him, were aware that enclosure was not limited to the eighteenth century, and to the Parliamentary movement (Leonard, 1905). On the other hand it was the intensity of enclosure which was striking in this period. It took centuries to enclose half of Leicestershire by 1730, but less than 100 years to complete the task.

Perhaps not surprisingly it was in areas which were already adapting to changing conditions by enclosing, that Parliamentary enclosure was most important, at least in its early stages. As a general rule the density of Parliamentary enclosure was greatest in the south and east Midlands, and progressively less important radiating outwards from that area. During the 1760s and 1770s the remaining heavier arable soils of the Midlands were enclosed by Act of Parliament, and large areas of arable were converted to grassland farming. Farmers and landlords were attempting to escape from the broadly fixed incomes they could obtain from open-field arable farming, and which could not be improved because little or no common waste survived for colonization. Consequently, the answer seemed to be to change the nature of farming by concentrating on animal husbandry. This was the case, for example, in the Leicestershire village of Wigston Magna (Hoskins, 1957). Although farm incomes recovered somewhat after 1750, the pattern was maintained of heavy-soiled regions moving into grassland farming, which is why so much of the early Parliamentary enclosure affected the south and east Midlands.

By the later 1790s the situation had changed considerably. In the French Revolutionary and Napoleonic wars between 1793 and 1815, food prices rose rapidly, and enclosing activity was designed to take advantage of this change. The intention was to improve the existing arable in light-soiled areas, and to extend arable cultivation in areas which were both economically and geographically marginal. Although Parliamentary enclosure had originally been inspired by the stresses and strains of open-field farming, the ease and success of the technique were now deployed to enclose waste and common; indeed, it is possible that in their fascination with the enclosure of the open fields historians have overlooked the fact that the general trend of enclosure may have been towards land reclamation rather than reorganization of the arable (Chapman, 1984, 1987). Be that as it may, from the start of the Napoleonic wars the waste was the principal target of the enclosers, including the extensive commons in areas such as Cumbria, regions which, with their leaning towards pasture farming, had escaped the worst effects of the price depression of the late seventeenth and early eighteenth centuries.

The long-term consequence was that 86 per cent of all Parliamentary enclosure took place by 1830 and most of the other 14 per cent by 1914. It was by no means uniform across the country. The Welsh borders, the south-east and south-west were hardly affected, whereas more than half of Oxfordshire, Northamptonshire and Cambridgeshire were enclosed by legislation. Even within a single county the variation could be considerable; in Buckinghamshire, 57.7 per cent of Cottesloe Hundred was enclosed by Act of Parliament, but just 7 per cent of Burnham Hundred (Turner, 1980).

Enclosure and the Landscape

Whatever the chronology of enclosure, the impact on the countryside was profound (Turner, 1984c). In open-field areas what had once been large, unhedged fields, were now divided into rigidly geometrical parcels, often of not much more than ten acres in extent, surrounded by fences or quickset hedges. The result was to alter the face of the countryside, and where land was turned down to grass it is still possible to see (although in fewer areas than was once the case) hedges cutting across the line of the ridge and furrow. Where waste and common were taken in, the appearance of earthen banks, or mile after mile of dry stone walling, had a similar impact. The walls went up even where the land use remained the same after enclosure, and they can still be seen snaking across the countryside over much of northern England.

Other changes in the landscape came about as forest and heaths were cleared for agricultural purposes. The Royal Forest of Delamere still covered nearly 10,000 acres of Cheshire in the early nineteenth century, despite the efforts of surrounding villages to enclose their common grazing land. The remaining forest was enclosed by Act of Parliament in 1812, and although large parts of it continued to be planted with timber, more than one third was granted to individuals who enclosed the land into large hedged fields. Heathlands were also divided into rectangular fields with earthen banks, and old woodlands were cleared to be made into neat oblong fields, usually as a result of an Act of Parliament. Marsh and fenlands were largely reclaimed (Taylor, 1975).

Enclosure and Farming Practice

The enclosure of communal land usually brought with it alterations to the whole farming operation, because the existing system was replaced by one in which hedges and fences separated holdings, and farming took place on an individual rather than a communal basis. The result was a change in the structure of both farming and landholdings, and it is usually argued that this was brought about in order to improve farming practices. In Ernle's words, 'where the soil was of a quality to respond quickly to turnips, clover, and artificial grasses, it was enclosed in order that it might profit by the new discoveries. This was the case on the light soils of Norfolk' (1961, p. 166).

There is a neatness about the assumption that open-field farming was necessarily bad, and that farming in enclosed fields was correspondingly good, which is perhaps questionable. In the immediate aftermath of enclosure, farming practices did not necessarily change. Enclosure could not compensate for the technical inefficiencies of farming, particularly as post-enclosure farmers were often the same people, with the same outlook, as had previously farmed in the open fields. In county Durham enclosure simply made more widespread the orthodox technique of two crops and a fallow (Macdonald, 1979b), while the land agent Thomas Davis commented of Wiltshire in 1811 that enclosure could bring improvement only if the farmers were ready and willing to innovate: 'though the common field husbandry doesn't make land better, it keeps it from becoming worse ... severalty makes a good farmer better, and a bad one worse' (quoted in Overton, 1983, p. 40). Furthermore, in some regions open-field farmers were able to adopt new crops within the open fields, hence the possible conclusion that the open fields were originally designed to maximize farming efficiency (Fenoaltea, 1988). Finally, as we have seen, in its initial stages enclosure was often brought about with the intention of changing the use to which land was put, either by converting open-field arable into pasture, or else by bringing into cultivation land previously in common or waste. Even in Norfolk it was recognized that improvement did not automatically follow from enclosure; in fact, it was more commonly found on land brought into cultivation (Parker, 1975).

If, however, enclosure made so little difference, why did landlords go to the trouble and expense of bringing it about? An obvious incentive was to raise the rent. Enclosure offered an opportunity to renegotiate lease agreements, and landlords almost certainly asked for a higher rent on an enclosed farm than they could obtain from open-field holdings. Some landlords even viewed enclosure in terms of the likely financial return. At Hibaldstow, in Lincolnshire, in 1795, detailed projections were made about the probable increase in rents to be expected from enclosure, while the evidence of five Nottinghamshire enclosures between 1787 and 1796 suggests that they were a response to anticipated monetary benefits (Beastall, 1978; Purdum, 1978). Nor were landlords disappointed. On Earl Fitzwilliam's estates in Northamptonshire and Huntingdonshire, even after taking into account the expense of new roads, fencing and hedging, the overall return in the seven cases for which acreages can be accurately calculated was 16 per cent (Thompson,

1963). Chambers and Mingay concluded that 'perhaps a doubling of rents, from about 7s to 15s per acre was the common result of enclosure in the Midlands' (1966, p. 85). Possibly this is an underestimate since rents trebled in Lincolnshire and Wiltshire (Grigg, 1966; Molland, 1959).

Rents alone are unlikely to offer a full explanation for enclosure, and we must also examine the significance of changes in farming practice. In the longer term it seems possible that enclosure brought about a structural change without which the widespread adoption of innovatory farming systems would almost certainly have been delayed if not prevented altogether. It also overcame the old limitations on the stock which farmers could keep. In the past these had been determined by the area of commons, meadow and pasture available.

Perhaps above all, however, enclosure gave farmers freedom of land use. In itself this was probably enough to persuade them to pay an increased rent since they gained sole rights over the land, which enabled them to search out the most profitable means of using it rather than being confined to the traditional rotation authorized by the manor court. This did not mean that farmers went over wholesale to new crops; instead they preferred to change the mixture by growing a greater variety. In Northamptonshire in 1801 enclosed parishes had a lower acreage than their open-field neighbours under pulses such as beans and peas, but a greater acreage under root crops. Fewer acres were probably cropped, especially for wheat and oats, but the land was better used, thereby releasing land for other crops and pasture. Farmers were also able to be more sensitive to market forces. Some of the new crops introduced on a large scale in post-enclosure Northamptonshire parishes point to an extension of meat and animal product farming, perhaps through conversion to pasture or the introduction of fodder crops into rotations. As a result, the same grain output could be achieved from a similar acreage, releasing land for more crops and animals (Turner, 1986b).

Ernle may have been right to suggest that the key to improved farming lay in enclosure but the mechanism was rather more complicated than he realized. If, in the longer term, it was re-sponsible for the structural changes which encouraged improved farming, in the immediate term its relationship with improved practices was probably variable and there is little direct evi-dence to suggest that enclosure was brought about primarily to promote innovation.

The social cost of Parliamentary enclosure in the countryside has usually been seen in terms of the dispossession of the small farmer from his acres and the cottager from his common rights. The people affected are thought to have left the land and headed for the towns, thereby providing the workforce required to operate machinery in the factories of the great new manufacturing centres of the Industrial Revolution. The argument has a long pedigree, and can be found in the work of historians with such differing political stances as Marx and Ernle (Marx, 1867, pp. 800−3, 829−30). Ernle wrote that 'hundreds of cottagers, deprived of the commons, experienced that lack of rural employment which drove them into the towns in search of work' (1961, p. 301).

By far the most forceful statement of the dispossession argument is to be found in J. L. and Barbara Hammond's book, *The Village Labourer, 1760−1832*. 'Enclosure,' they wrote, 'was fatal to three classes: the small farmer, the cottager, and the squatter. To all these classes their common rights were worth more than anything they received in return.' Small farmers received land at enclosure, but most were 'overwhelmed' by the legal costs and expenses of fencing, and forced to sell up. For those who rode these costs, the loss of fallow and stubble grazing was sufficient to pull the remaining carpet from under their feet: 'the small farmer either emigrated to America or to an industrial town, or became a day labourer.' In other words, the displaced small farmers formed the basis of the new proletariat. The cottagers fared no better: 'the effect on the cottager can best be described by saying that before enclosure the cottager was a labourer with land, after enclosure he was a labourer without land. The economic basis of his independence was destroyed.' The loss of traditional right to cut furze and turf from the common, and of the prescriptive right of keeping a cow, weakened their position. Even those who owned their cottages were compensated with 'something infinitely less valuable ... a tiny allotment ... worth much less than a common right'. Squatters, people who had 'settled on a waste, built a cottage, and got together a few geese or sheep, perhaps even a horse or a cow, and proceeded to cultivate the ground', also suffered. Those who had lived in the parish for 20 years or more often received some form of compensation at enclosure, but like the cottagers they lost their common rights (1911, pp. 97−103).

J. L. and Barbara Hammond were not alone in this view. Gilbert Slater had taken a similar line in 1907; George Bourne, writing in 1912, made stirring references to the damage done by the process of enclosure; and E. C. K. Gonner, also writing in 1912, expressed the view that as a result of enclosure 'there is little room for wonder at the steady and widespread disappearance of the small owner cultivating his own little farm' (1912, p. 369). The dispossession argument became the conventional wisdom.

Few historians have been prepared to question the assumption that Parliamentary enclosure represented nothing less than class robbery. The large and powerful landowners were presented with a means of legally dispossessing their lesser brethren, and they were more than willing to exploit their contacts with Members of Parliament and peers to ensure a smooth passage for their (private) legislation (Martin, 1979b). Although it has sometimes been argued that relatively little opposition was mounted to enclosure proceedings, this may be a mistaken view. Writing about the benefits of enclosing the manor of Loughborough in Leicestershire in 1744, William Gardiner told the Earl of Huntingdon that: 'the farmers there will all be against it, as they know it must cause some of them to seek for farms in other places, because when enclosed one man that has substance can with the same number of servants manage a farm of three, four, or six times the value' (Historical Manuscripts Commission, *Hastings MSS*, vol. III, 1934, 47). Opposition was mounted to more or less every stage of Parliamentary enclosure in Northamptonshire, particularly in villages with significant numbers of artisans and mechanics (Neeson, 1984), but it tended to be passive rather than active in Buckinghamshire (Turner, 1988). Nationally, 36 per cent of all enclosure bills ran into some kind of dissent during their progress. The level of opposition mounted to Parliamentary enclosure suggests that many groups believed themselves to be penalized.

The dispossession argument must not, however, be accepted uncritically. First, the enclosure movement did not affect the whole country; in fact, only one in four or five of the rural population was directly affected. Moreover, the problem of rural poverty was particularly bad in southern England where enclosing activity was rather less pronounced than in the Midlands and north, and where the opportunities for employment in nearby industrial towns were fewest. Second, none of the three groups identified by J. L. and Barbara Hammond as the losers from enclosure seem to have been as badly hit as might be expected from the emotive language. Among small farmers in the south Midlands Parliamentary enclosure seems to have had little impact on numbers, and those who did leave the land were often replaced by men of similar standing and acreage (Allen, 1986a; Turner, 1975).

42

As for cottagers and squatters, their decline also needs to be kept in perspective. The extent to which they still possessed viable common rights on the eve of enclosure has been debated, while there is little doubt that enclosure increased, rather than decreased, their employment opportunities. Of course, it would be pointless to argue that nobody left the land as a result of enclosure. In Buckinghamshire, common-right owners seem to have been the first casualties of land sales after enclosure. Even if such individuals reinvested their capital in a tenancy, this was occurring when farms were growing in size, which would have made life difficult for them. Some of those selling must have become completely landless or else they replaced existing tenants. The process is unclear but there could have been a squeezing at the bottom end of the agricultural ladder leading to landlessness and difficult conditions for the agricultural labour force (Snell, 1985).

The most compelling evidence against declining numbers, however, concerns employment opportunities after enclosure, and of a post-enclosure rural population which continued to grow at least until the 1840s. Various reasons can be suggested as to why this was the case, including the number of jobs created at enclosure and the extension of labour-intensive cropping such as turnip growing. This is not to deny either that the farming community was releasing labour on a significant scale, since the proportion of the total workforce employed on the land approximately halved between 1760 and 1840, or that enclosure tended to increase some types of unemployment. Particularly in southern England the gradual decline of service in husbandry for young males aged 15–24 meant that instead of being protected from poverty through the winter months by their annual contracts they were turned out to depend on poor relief (Kussmaul, 1981). The decline of employment opportunities for female labour may also have had the result of cutting living standards.

Part of the difficulty is to distinguish between the effects of enclosure, and the problems arising from changes in agricultural practice and the economic reversals of the post-1815 years. By the 1830s, particularly in southern England, the agricultural labour force had been to a considerable extent proletarianized. In Norfolk in 1816 food and wage riots were followed by machine breaking as a result of falling wages in the difficult economic climate at the end of the war. Further troubles followed in 1822, and rural uprisings took place throughout southern England during the so-called Captain Swing riots of 1830. Rural discontent was also fuelled in other ways. One was the so-called scandal of open and

43

closed villages. Labourers were excluded from 'closed' villages to keep down the poor rate, and forced as a result to live in cramped conditions in 'open' villages, although the true extent of this practice has been questioned (Banks, 1988).

Taken together, the consequences of these trends was that for the rural labour force life may have been rather more precariously balanced in financial terms after 1815 than previously. Many ended up claiming poor relief for at least part of the year, and the problem was especially marked in the rural areas furthest removed from the new industrial centres. On the other hand, the distribution of poverty may have had less to do with enclosure than with the availability of work outside of farming. Be that as it may, the traditional view of small farmers, cottagers and squatters being collectively dispossessed of their rights and driven into the new factory towns can no longer be fully maintained (Armstrong, 1988).

4 Land

Although innovation and enclosure have been regarded as the crucial factors promoting the Agricultural Revolution, the context of landownership and farm sizes in which they occurred was also vital. It was changes in the agrarian structure which first attracted attention to the Agricultural Revolution, and it remains an issue of considerable importance. It has usually been argued that by 1750 English land was held in substantial estates by landlords who adopted a capitalist outlook and let their land in large farms for the highest possible rents. In the words of Ernle, 'it was the large landlords who took the lead in the agricultural revolution of the eighteenth century, and the larger farmers who were the first to adopt improvements. Both classes found that land was the most profitable investment for their capital' (1961, p. 161). Most recently this enduring argument has been repeated by Brenner in his search for an explanation of why English farmers were so much more successful than their European, and particularly French, counterparts in the eighteenth and nineteenth centuries.

In Brenner's view the relatively harmonious relations between English landlords and the monarchy ensured a community of interests and not a conflict (as in France). This enabled English landlords to consolidate their properties as and when conditions were favourable. By the late seventeenth century the larger estate owners controlled 70–5 per cent of cultivated land. Such a development stimulated economic growth because landlords created large farms which they leased to capitalist tenants who could afford to make the necessary investments and who drew on a free labour market. Since neither rents nor dues were excessive, improvement could and did occur. The growth of agricultural productivity was therefore rooted in the transformation of agrarian class or property relationships, and this created the conditions in which a dynamic internal economy pushed the country towards industrialization (Brenner, 1976). Brenner has been criticized, but in a reply to his critics he pointed out that he drew a contrast between English and Continental development because England was 'marked by the rise of a *capitalist aristocracy* which was presiding over an *agricultural revolution*' (Brenner, 1982, p. 89). The implication is that whatever the role of innovation, Marx was essentially right to stress the importance of changes in the agrarian structure. Nor has this emphasis been disputed by historians of a different outlook. Chambers and Mingay wrote in 1966 that 'at its best the English landlord-tenant system was reasonably efficient and flexible — far more so than the conservative peasant farming of the continent — and it provided the essential framework for the great leap forward of agriculture in the eighteenth and nineteenth centuries' (1966, p. 21). Consequently we must examine the various strands of the argument.

The Rise of the Large Estate

In late nineteenth century England and Wales the vast majority of the land was held in extensive estates. For commentators such as de Tocqueville and James Caird, this was a natural result of agricultural and industrial change. Others argued that the trend had begun much earlier, primarily in connection with changes in the land laws and inheritance practices dating from the mid-seventeenth century. This view, that the greater estates were built up at the expense of smaller owners during the seventeenth and eighteenth centuries, was developed by Marx, and followed in a number of other studies. In more recent times Habakkuk narrowed the timing down to the 50 or so years after the Glorious Revolution of 1688, a period when agricultural prices, heavy taxation and the implementation of legal changes (in particular the so-called 'strict settlement'), collectively served to restrict the supply of land. By the mid-eighteenth century a nation of large estates and tenant farmers had emerged (Habakkuk, 1939—40). We now know that this rather stark view of events oversimplifies changes which took centuries rather than decades, and which were marked by considerable regional variation. Studies of different areas have revealed a variety of consolidation patterns between 1660 and the late nineteenth century. Consequently, although there is little doubt that land was consolidated into larger estates, this process was certainly not completed by 1750—80 as was once believed (Beckett, 1977; 1984).

According to the traditional argument, from 1660, if not earlier, the small landowner, usually an owner−occupier working predominantly his own land, was gradually dispossessed of his holding and turned into a tenant farmer or wage labourer. Owner−occupiers are thought to have owned around 33 per cent of English acreage in the later seventeenth century, but a mere 10−12 per cent by the 1870s. Their demise is usually dated to coincide with the accumulation of large estates and the efforts of landlords to reduce the number of small − and presumably inefficient − farmers, hence the conclusion that by the second half of the eighteenth century capitalist class relations predominated in the countryside. Territorial magnates let their property to substantial tenants employing wage labour, while owner−occupiers were reduced to an insignificant rump.

The demise of the owner−occupier by *c*.1750 is now well established in the literature. Marx argued that the traditional English 'yeoman' was gradually dispossessed of his rights over the land during the period 1450−1750. He became instead a tenant farmer, or he left the land altogether. The argument was developed by various historians (Johnson, 1909; Habakkuk, 1965), and Mingay has concluded that 'a major decline must have occurred in the hundred years after 1688 ... the small owner−occupiers had only 10−12 per cent of the cultivated acreage in the later nineteenth century, and a not very much larger percentage in the late eighteenth century' (1968, pp. 15−16). There have been dissenting voices from this consensus, but the general view is that derived from Marx.

However, the consensus raises a number of questions given the increasing uncertainty about the timing of great estate consolidation. Contemporaries seem to have been by no means as convinced as historians about the speed of decline. Patrick Colquhoun's survey of English society at the beginning of the nineteenth century found no demise of the freeholder, and recent attempts to revise his figures actually postulate a rise in numbers (Lindert and Williamson, 1982). Owner−occupiers survived in many areas, not all of them on the peripheries where agricultural difficulties were felt less severely and where enclosure was less significant. Despite enclosures, in mid-eighteenth-century Leicestershire owners of 100 acres or less 'remained by the middle of the eighteenth century, collectively at least, as great in landowning strength as any other single class in many parishes' (Hunt, 1958−9, p. 501). Even after enclosure small owners were not obliterated in counties such as Buckinghamshire and Warwickshire (Martin, 1979a), and nor was the picture particularly different outside the claylands (Beckett, n.d.). In fact the decline may have been exaggerated because historians have based their findings on the land tax assessments, which are now known to be unreliable in this context.

Two points in particular emerge from the mass of evidence which has been collected. First, enclosure may have brought a considerable turnover in landowning personnel, but this is not the same as the disappearance of small owners. Second, there is little doubt that many small parcels of land changed hands during the war years with an upsurge in the number of owner−occupiers. Together these points suggest that during the war years the number of owner−occupiers increased but the personnel changed. In Buckinghamshire, for example, the rate of turnover of families within a few years of enclosure was more than 30 per cent of the original owners, and the norm was 40−50 per cent (Turner, 1975). Many came to grief in the years immediately after Waterloo, and those who survived declined slowly in the course of the nineteenth century. Regional differences were important, since the smaller owners fared better in the north and west than they did in the south and Midlands. A further important change was the shift away from owner-occupancy towards holding accommodation land. By the 1870s nearly 25 per cent of the landed acreage was in the hands of individuals owning 300 acres or less, but many of these people had no attachment to the land as farmers. Many had purchased land as an investment, leaving only about 10 per cent of the land still owner−occupied in the true sense of the word (Reed, M., 1984).

Farm Sizes

If the consolidation of the larger landed estates took longer to bring about than was once believed, did this also mean that the growth of farms was correspondingly delayed? Large farms run by capitalist − sometimes 'gentlemen' − farmers employing waged labour have long been regarded as a vital prerequisite for the Agricultural Revolution. According to Ernle:

the new system of farming required large holdings, to which a new class of tenant of superior education and intelligence was attracted. It was on these holdings that capital could be expended to the greatest advantage, that meat and corn could be grown in the largest quantities, that most use could be made of those mechanical aids which cheapened production. (1961, p. 215)

The tenants of large farms, who had to be men of sound financial standing, were likely to practice the most improved farming and thereby to achieve higher yields than on small farms. In addition, they had greater bargaining power in the market place, they were more efficient in the use of labour and capital, and they were able to hire the best labourers and to make the most economical use of implements. Arthur Young had nothing but praise for large farms, 'small farms are detrimental to the occupier and public in the smallness of their produce, rather injurious than otherwise to population . . . large farms, in respect of produce are more beneficial than any — more advantageous than any to population in number and the value of land' (Young, 1767). For Ernle their virtues were found in most abundance at Holkhan: 'in Thomas Coke the new system of large farms and large capital found their most celebrated champion . . . Coke stimulated the enterprise of his tenants, encouraged them to put more money and more labour into the land, and assisted them to take advantage of every new invention and discovery' (1961, pp. 217, 219).

These panegyrics raise three questions: what was a large farm; were farms increasing in size; and how important were large farms in the Agricultural Revolution? The answer to the first question is more complicated than we might anticipate. Contemporaries frequently used the terms large and small in relation to farms, but they seldom defined what they meant. Arthur Young, who once noted that 'one can scarce ever be accurate in using such terms as *large* and *small*', seems to have regarded the minimum size of a large farm as 300 acres, but among his contemporaries Thomas Batchelor considered it to be 200 acres, and William Marshall referred in 1777 to 100—500 acres as 'middling' and in 1796 to 100—300 acres as 'the middle cast'. We may take it that the minimum size of a 'large' farm cannot have been less than 100 acres. Certainly Young could not conceive of any good farms being smaller than this (Beckett, 1983a).

Second, were farms increasing in size? R. H. Tawney dated the rise of the large estate and the large farm to the sixteenth century, but more recent research points towards an explosion of small cultivators, hence Outhwaite's conclusion that there is 'a strong probability that the number of small farms was growing rather than declining' (Outhwaite, 1986). For the eighteenth century, examples of farm size increases abound (Mingay, 1961—2). Between 1714 and 1832, on the Leveson-Gower estates in Staffordshire and Shropshire, the average size of farms over 20 acres increased from 83 to 147 acres (Wordie, 1974). Allen has argued that farm sizes increased substantially on estates in the south Midlands except in open-field areas where increases did not occur before the middle of the eighteenth century. Whereas in the early seventeenth century just 32 per cent of open-field land was held in farms of 100 acres or more, by 1800 85 per cent of the open land was in such farms (Allen, 1988). The picture was similar in Laxton where farms of 100 acres or more accounted for 32 per cent of the parish acreage in 1635, 44 per cent in 1736, 51 per cent by 1789, and 60 per cent by 1820. Farm sizes also grew steadily through the seventeenth century in the Leicestershire village of Wigston Magna (Hoskins, 1957). On the other hand, the pattern was not found everywhere; little radical change occurred in south Lincolnshire farm sizes between 1770 and 1850 (Grigg, 1966).

The gradual increase in farm sizes through the eighteenth and into the nineteenth century is not in question, but the pace and extent of change needs to be looked at carefully. Reporters to the Board of Agriculture around 1800 noted engrossment in a number of counties including Bedfordshire, Cheshire, Devon, Dorset, Shropshire, Staffordshire and Wiltshire, but the continued importance of small farms in no fewer than sixteen counties, among them Nottinghamshire and Northamptonshire (Marshall, 1818). For this, various reasons were adduced, among them the persistence of open-field farming, tithes and the absence of leases. A further reason which we might offer would be farming conditions. The decline of the small family farmer varied across the country. In both Wiltshire and Cambridgeshire for example it occurred most rapidly on the chalklands. In addition it was not a consistent trend, since there is plenty of evidence to suggest a resurgence of small farms during the Napoleonic war years, partly due to the increase in owner-occupiers (Molland, 1959).

Contemporaries and historians alike have suggested that innovation occurred on large farms, and that the land for such farms was made available by enclosure. Thus, large farms resulted from a combination of the disappropriation of the owner-occupier and the small tenant farmer, and the process of Parliamentary enclosure. As Thomas Wright wrote in 1795, enclosures were 'injurious to the peasantry: for the destruction of the peasant's cottage, whereby he loses the opportunity of raising stock on the waste, may operate in the same manner as the monopoly of small farms' (Wright, 1795, p. 15). Hence arises the argument that one of the incentives to enclose was the desire of landlords to turn their estates into more economic farms deploying up-to-date farming practices and yielding high rents. This in turn has fostered a belief that the farming innovations were essentially a response to large, post-enclosure farms. The clearest statement of this viewpoint was made by Herman Levy, who suggested that between 1750 and 1850 the small farm was squeezed out by a combination of enclosure and the movement of grain prices in relation to livestock prices (Levy, 1911).

There are several problems with this argument. Land released at enclosure was not automatically reallocated into large farms. John Donaldson, reporting from Northamptonshire to the Board of Agriculture in 1794, noted that following enclosure land had generally been let out in small parcels to existing tenants. In addition, farm engrossment continued in areas of early enclosure, there was no automatic correlation between the decline of the small farm and the Parliamentary enclosure movement, and nor is there much evidence to support the view that land released to larger owners by the decline of the owner−occupier was automatically reorganized into large farms (Saville, 1969). This may have happened, but it remains to be convincingly demonstrated. Perhaps more significantly, in the south Midlands farm sizes grew during the eighteenth century in both open-field and enclosed parishes.

The trend towards consolidation continued in the nineteenth century, but not to the exclusion of the smaller farmer. In 1851 just over half the farmed land of England and Wales was laid out in holdings of 200 or more acres, and four fifths was in farms exceeding 100 acres. These figures may well understate the number of smaller farms (Grigg, 1987), and as James Caird became aware when he toured the country in 1850−1 large and small, progressive and antiquated farms were often found in adjacent areas. He noted a distinctive pattern of farm sizes:

> in the dry climate of the counties on the east coast, the operations of a corn farm can be carried on, with great precision and regularity on an extensive scale … as we proceed westward, the country becomes more wooded, and better adapted for pasturage; the enclosures are smaller, the farms less extensive, and the farmers more numerous. (1852, pp. 481−2)

The largest farms were found in the south Midlands, East Anglia and the southern counties, while much smaller farms predominated in areas such as the north Midlands and the south-west (Grigg, 1963). In fact the 20 per cent or so of the cultivated acreage in farms of fewer than 100 acres was still highly fragmented in the late nineteenth century. In 1878 Caird calculated that only 18 per cent of tenanted farms exceeded 100 acres, and that 70 per cent were of 50 acres of less (Caird, 1878). Nearly a decade later Craigie concluded that for a country usually described as one of larger farmers, England had a surprisingly high proportion of small holdings. He found that 71 per cent of holdings were under 50 acres, with barely 1 in 100 topping 500 acres (Craigie, 1887).

The final question is: did large farms affect farming practice and did the speed of growth determine the pace of innovation? The problem is to distinguish between theory and practice. Arthur Young never ceased to extol the virtues of large farms, but a reworking of his data suggests that large and small farmers were not substantially different in the capital they employed per acre, that there is relatively little evidence for larger farmers being more innovative than small farmers, and, perhaps as a result, large farms were not necessarily higher yielding (Allen, 1986b). Although Coke's farms were impressive it has to be remembered that they were already large before he became the landlord, and they were almost certainly not typical (Parker, 1975).

We need therefore to look beyond Arthur Young when considering the virtues of large farms, and there is plenty of evidence to suggest that contemporaries regarded them as a mixed blessing. Nathaniel Kent, William Marshall, and a number of other respected agricultural commentators argued persuasively that whatever the merits of large farms, the ideal from a farming point of view was a mixture determined by factors such as the availability of capital and the nature of farming. Capital requirements depended on farm sizes, and contemporaries believed that one of the major reasons why farms did not grow larger was the shortfall of suitable tenants. As Thomas Batchelor commented of Bedfordshire in 1808, farms had remained small on the poor clay soils north of the river Ouse, 'which affords no great temptation to opulent farmers' (Batchelor, 1808, p. 27).

In practice the transition to large farms was neither as complete as the arguments about agrarian capitalism would imply, nor as straightforward as Arthur Young would have liked. No one would deny that in the longer term farm sizes did increase, but many small farms remained at the end of the nineteenth century, and the correlation between enclosure and farm size increase remains uncertain.

5 Output

So far we have examined the processes of change which constituted the Agricultural Revolution. Now we must take a more detailed look at the impact they had on agricultural output. Assumptions about output are often based on the evidence of population growth. They may well be justified, but they are negative measures of an important change. Output has to be measured in terms of percentage production increase over a period of time. It can be raised in one of two ways; by extending the acreage under cultivation, or by improving the output of the land already in use. We have looked at the efforts made to increase the cultivated acreage, and now we must turn to the question of how far innovation, enclosure and farm sizes helped to raise output. By examining trends in output and productivity we can then assess the contribution of agriculture to the growth of the economy (chapter 6).

The task of estimating output is complex. In 1866 the Board of Agriculture began to collect and publish annual output statistics. Before that time the statistical data is thin, and regional variation was such that any gross national figures have to be treated with considerable caution. Even so, there can be little doubt that the overall output of English agriculture rose substantially over time. Gross cereal output increased between 1695 and 1750 by approximately 19 per cent (Chartres, 1984—5). The biggest increase was in wheat production which nearly doubled over the period, mainly at the expense of rye and to some extent barley. This trend continued after 1750. Between 1750 and the 1840s the acreage under wheat at least doubled, largely before 1812 and at the expense of rye and barley. The best available estimates suggest that the output of wheat increased by approximately 74 per cent during the eighteenth century. Output may have doubled between c.1800 and the early 1820s with another 50 per cent increase to the mid-1830s. From 3.175 million quarters in 1700 output reached 18.665 million in 1845, an increase of nearly 500 per cent. Recent figures suggest increases in output during the century 1750−1850 of 225 per cent in wheat, 68 per cent in barley and 65 per cent in oats (Holderness, 1989).

The output of meat can be calculated only by taking into account both the number of animals slaughtered, and changes in their weight over time. Contemporaries were vague when it came to estimating output. One suggestion, made in the early years of the nineteenth century, was that 'the size and weight, both of cattle and sheep, have probably increased at least one-fourth since 1732' (General Report, 1808, p. 43). Figures for Smithfield market show a 220 per cent increase in the number of cattle brought to market over the period 1750−1850, and an increase of 135 per cent in sheep. An increase in output from slightly less than 6 million hundredweight a year in 1750 to over 12 million by 1850 is not out of the question and may indeed be an underestimate since it suggests a decline in per capita consumption. Wool output also doubled between 1750 and 1850 (Mingay, 1989, pp. 1063−4).

Overall, a percentage increase in total agricultural output during the eighteenth century approximately double the level achieved in the two preceding centuries is not unreasonable, hence Holderness's conclusion that between 1750 and 1850 the general output of English agriculture 'rather more than doubled' (Holderness, 1989, p. 174). After 1850, and despite the prolonged depression in agriculture from 1873, output continued to grow. Recent estimates suggest a rate of 0.3 per cent per annum from 1871 to 1911, which was rather less than the British economy as a whole achieved (Ó Gráda, 1981, p. 178).

Land Use

To what extent were these increases achieved by extending the cultivated acreage, and how far did they depend on better use of the land already in production? These questions are materially different but it is not easy to draw a clear distinction between productivity increases resulting from an extension of the acreage and increases arising from improved methods. Increases in output were partly achieved by taking in land for cultivation. Even in East Anglia output increases between 1650 and 1750 may have reflected more land being cultivated rather than rising yields (Overton, 1983). Nationally there was a considerable increase in the cropped acreage during the Agricultural Revolution. Between 1695 and 1750 the acreage under cereals in England and Wales increased from 5.375 million to 5.732 million. In 1750 between 9.5 and 10.5 million acres were under crops or fallows. This figure rose to about 12 million acres by the end of the Napoleonic wars, and by 1850 the total was probably about 13.5 million or a little more (Holderness, 1989, pp. 126–7).

Obviously a significant proportion of the rise in output was achieved by extending the cropped acreage, but we must not assume that the yields obtained from new land were comparable with land already in use. 'New' was in many cases marginal land, and while it was rather more productive under the plough than as waste or moor, it was likely to produce poorer yields than on the better land already in use. Even with improved rotations the return on investment was likely to be a short-term one, as in the case of uplands and hillsides cropped for a few years during the Napoleonic war years. Moreover, when the lower yields achieved on these marginal lands are incorporated into average yields the overall figure is likely to be reduced. In this way the cultivation of new land can give the impression of lowering yields.

Land and Productivity

The increase in cultivated land was considerable but it cannot explain all the rise in output. Between 1695 and 1750, for example, the gross output of cereals grew by 19 per cent although the area of land sown increased by just under 7 per cent. Similar figures could be quoted for later periods, but the implication is that changes in farming practices were having a knock-on effect on productivity. This is arable output, and unfortunately we know far less about livestock productivity.

Calculations of land productivity are fraught with difficulty, partly because the data tends to be scattered and inconsistent. However, some broad trends are accepted. Hoskins argued that yield ratios (the relationship between seed sown and grain harvested) roughly doubled between about 1500 and 1650, but that there was no further discernible rise between 1650 and 1800 (Hoskins, 1968). This may be too optimistic, at least at the national level (Outhwaite, 1986). For the eighteenth century most historians accept Deane and Cole's estimate of a 10 per cent improvement in wheat yields (the output of grain per acre), although few have been convinced by their methodology. In an effort to find more acceptable statistics the question of productivity has recently been approached from a different direction, using the evidence of probate inventories and of the agricultural returns prepared for the government during the 1790s.

Estimates based on Norfolk and Suffolk probate inventories suggest that wheat yields may have increased by 25–40 per cent, from about 8 bushels an acre in the 1580s to 14 by the early eighteenth century. Nationally, yields probably touched about 16 in 1750 rising to in the region of 30 by the 1830s (Jones, 1974; Overton, 1979). On the basis of the 1790s crop returns Turner has argued that productivity change was much in excess of 10 per cent, and the period of greatest change was before 1770 (Turner, 1982; 1984b). However, the most recent estimates suggest that wheat yields rose from about 18 bushels per acre in the period 1750–70, to 21.5 bushels per acre for 1795–1800, to 23 bushels per acre in 1810, and to 28 bushels by 1851.

By the 1840s yields of wheat were consistently higher than in the past as a result of the application of artificial fertilizers, improved drainage on the clays, and a better quality of dung derived from cattle fed on purchased feeding stuffs. Figures for wheat yields in the 1840s and 1850s show them to have been consistently higher than those for the 1820s and 1830s. However, figures produced by Healy and Jones (1962) overstate the case, and Holderness has concluded that the main achievement of the period 1800–50 was in bringing average husbandry up to a standard approaching that of the best practitiqners (Holderness, 1989).

These increases were partly achieved by the introduction of new crops. It is possible to make at least tentative estimates of the improvements in agriculture as a result of increased nitrogen supply once legumes replaced fallow, and the effect that this had on the output of grain (Chorley, 1981). However, productivity cannot be measured purely in terms of the role of clover and turnips. In Hertfordshire output per acre seems to have increased in the later seventeenth century partly through farm size increase, and partly through a shift away from lower yielding and lower quality crops like rye, towards higher yielding grains like barley. Systematic ground and seed-bed preparation seems to have been the key to improving yields in the county, and the introduction of fodder crops was not a necessary condition for sustained improvements (Glennie, 1988b). Increases in output also varied according to soils and farming interests. Rapid increases were probably achieved in regions where innovations were significant. In areas such as East Anglia they may have gone up by something in the region of 75 per cent, whereas on the traditional clayland arable 5 per cent may have been a good return.

The increases must also have had something to do with enclosure, but on this there is less agreement. Allen, drawing on evidence from the south Midlands, has argued that yields rose through the seventeenth and eighteenth centuries. Data from Oxfordshire suggest that there was little change before 1600, but considerable increase during the seventeenth century. It follows that most of the yield increases achieved by south Midlands' farmers between the Middle Ages and the nineteenth century was achieved by the end of the seventeenth century. The drawback to this viewpoint is that if so much had been achieved by 1700 it must have occurred within the context of open-field farming. Allen accepts that the logic of his argument is that much of the increase was brought about by open-field farmers in the late seventeenth and early eighteenth centuries. Used together, he argues that rent and probate evidence support the view that considerable technical change took place during this period in the open fields, with the result that corn yields virtually doubled in the Midland counties between the Middle Ages and the nineteenth century (Allen, 1986b). Unfortunately it is not entirely clear from Allen's evidence how this was achieved, although it is well known that innovation occurred in open fields in some parts of Oxfordshire (Havinden, 1961). In practice, Oxfordshire may be an unfortunate example to chose since in 1830 it retained a considerable proportion of its open fields — perhaps reflecting a level of local contentment with the existing situation which was not shared elsewhere.

Contemporaries were in little doubt that enclosure was the key element in raising productivity. Arthur Young, ever the optimist, never ceased to regale his readers with stories of major increases in output, and to compare them with the uncertain yields achieved in open fields. Moreover, rising rents could easily be demonstrated, and these appeared to prove the point. There is obvious logic in using rents as a benchmark. Productivity is difficult to measure, so rental improvements can act as a surrogate for productivity on the grounds that rents could only increase if farmers could afford to pay them. Improved rents can then be seen as an indication of improved yields (Turner, 1986b). The difficulty is to distinguish rent increases as a result of enclosure from what was a general upward movement of rents from about 1750, and a particularly rapid rise 1793—1815.

In themselves, therefore, rents are not an adequate measure of gains in productivity brought about by enclosure. After all enclosure could produce direct gains simply by eliminating losses through trespass — in the open fields farmers crossed other people's land to reach their own − too frequent fallowing in the fixed rotation of the open fields, and the probable inefficiency inherent in a system of scattered holdings. Furthermore, improved yields did not even require the introduction of new crops. A simple change in cropping procedures may have been sufficient to raise output quite significantly, particularly where the most suitable crops were grown on the most suitable soils.

More difficult to estimate is how far post-enclosure productivity gains resulted from the introduction of actual methods of improvement. In some parts of the country there is evidence of changing practice after enclosure, presumably designed to increase output. In Northamptonshire in 1801 enclosed farms achieved higher yields in terms of grain output per acre than those still in open fields, and this was more than sufficient to compensate for the lower acreage under grain. On 5.6 per cent less grain land farmers on enclosed farms produced 10 per cent more bushels of grain than their open-field counterparts, thereby increasing output and releasing land for other uses. As a result, more grain was produced than in the open fields and the land saved could be used for other purposes. However, Northamptonshire was just one county, and it seems likely that the situation differed regionally (Turner, 1982; 1986b).

We cannot therefore assume a straightforward correlation between enclosure and improved farming practices, and perhaps as a result historians have differed about the consequences of enclosure for productivity. First there are the optimists, summed up perhaps by Wordie's conclusion that:

the gains derived from enclosure in terms of productivity levels per acre of land were somewhere between substantial and formidable ... the output gains in terms of the cash value of produce may have been anything between 50 per cent and 100 per cent, once all the technical advantages available to the enclosed farmer had been fully deployed. (Wordie, 1983, pp. 504−5)

Second come those who take a less positive stance. Thus McCloskey has argued that 'a village was roughly 13 per cent more productive in an enclosed than in an open state' (McCloskey, 1975, p. 160). While this may be true of the difference between well-cultivated, open-field arable before and after enclosure, the yield on common grazing land, and on under or unused pasture, may have been considerably greater. Finally come the pessimists. Allen, for example, has argued that contemporaries, and particularly the irrepressible Arthur Young, may have exaggerated the extent to which enclosure raised yields, and that productivity calculations do not suggest that enclosed farms were any better farmed than those in the open fields (Allen, 1982; Allen and Ó Gráda, 1988; Fenoaltea, 1988).

Labour and Productivity

We cannot leave the question of productivity without examining the role of labour. One of the most remarkable features of the Agricultural Revolution was the growth in labour productivity relative to other European states. In 1500 labour productivity was relatively similar in England and France, but between 1600 and 1800 output per worker grew by 73 per cent in England but only 17 per cent in France (Wrigley, 1985). Indirect estimates suggest that output per person could have increased over the century 1650–1750 by rather more than one third, and that the pace of change accelerated in the century which followed. Jones has calculated at 47 per cent increase in labour productivity during the eighteenth century, and has also suggested that the labour market became more sensitive to employers' demands (Jones, 1981). By the early nineteenth century labour productivity was much higher in the United Kingdom than elsewhere in Europe (O'Brien and Keyder, 1978). In 1760 each agricultural worker's output was capable of feeding about one other person; by 1841 the ratio had risen to 2.7 non-agricultural workers. Over the period 1700–1850 the agricultural labour force probably increased by between 50–75 per cent, and labour productivity rose very considerably given that the volume of output in the mid-nineteenth century was more than three times the level achieved in 1700. This suggests that labour productivity rose at approximately 0.5 per cent per annum over the period. Such levels may have given English agriculture an important boost by comparison with her European neighbours.

The question is, how were these improvements achieved since, as we have already seen, machinery (with the exception of threshing machines) was of very little importance before the mid-nineteenth century? The simple answer is that English labourers must have worked harder than their Continental counterparts. There is some evidence to support this contention (Clark, 1987), although contemporaries were not so sure. François de la Rochefoucauld maintained that French agricultural labourers in the 1780s worked 'much longer hours, starting an hour earlier in summer and finishing nearly two hours later'. He found Suffolk labourers worked 'very casually, resting often and conversing a lot. I am certain that a French day worker does almost a fifth more work in a day than an Englishman' (Scarfe, 1988, pp. 54–5). Moreover, English farmers often complained about the quality of their labourers (Armstrong, 1988). Allen has argued that it was the growth of farm sizes which was critical. He has questioned the views of Arthur Young in the eighteenth century, and J. D. Chambers more recently, who argued that large farms increased both employment and output. Instead, he suggests that employment declined thus raising labour productivity (Allen, 1988).

At present there is no single answer to this particular problem. Labour productivity probably increased for a variety of reasons, among them regional specialization in farming (barley in Norfolk and pasture in the Midlands, for example), capital investment in the form of enclosure and drainage, and the substitution of animal for human power. The introduction of fodder crops increased the capacity of farmland to carry more animals with the result that by the early nineteenth century (if not earlier) a major distinction between England and much of Europe was the more animal-intensive character of its farming (O'Brien, 1985). This may be the key to understanding the growth of labour productivity, but it remains to unlock the door.

6 Agriculture and the Economy

Traditionally the Agricultural Revolution is held to have occurred in tandem with the Industrial Revolution. The agricultural sector supplied raw materials and food, and released labour – in the sense of a relative and not an absolute decline in the number of farm workers as a proportion of all labourers – as well as generating investable funds and the taxes needed to sustain an industrial and urban population. The rise in incomes in terms of rents and profits as a result of the growth of agricultural productivity enabled consumers to purchase larger quantities of manufactured goods and urban services available on the home market. This in turn stimulated industrial production. Thus the two movements were intimately related while the great boosts to the industrial economy after 1830 were seen to be paralleled in agriculture by High Farming. It was only in the 1870s and beyond that agriculture fell behind, partly as a result of overseas competition. As we have already seen many of the assumptions about agriculture on which this interpretation is based no longer hold good. Consequently we need to ask what was the role of the farming community in the development of the economy, and in particular did it retard or promote growth?

From the 1540s a positive relationship existed between the rate of population growth and the rate of growth in food prices, except for price downturns in the immediate aftermath of wars. Consequently, despite a downturn in both population and prices output continued to grow quickly *c.*1650–1740 as a combination of comparative plenty and falling prices stimulated organizational changes and hastened the introduction of new techniques. The evidence suggests continuing productivity gains through this period, and overall an impressive performance by British agriculture in the years before 1760. Growth, according to Crafts, ran at 6.2 per cent per decade over the period 1700–60. Usually this is seen as overproduction only partially offset by exports so that through the period 1710−45 the relative price of agricultural goods declined (Crafts, 1985).

What was the impact on the economy? For the century or so before 1750 historians have emphasized the 'centrality of agricultural change and a dynamic home market to British economic growth'. The impact of grain prices on consumption is not easy to measure, but A. H. John concluded that the changes of the 1660−1750 period, and particularly the evidence of depression during the 1730s and 1740s, were of vital importance to the economy: 'The long-term fall in bread-grain prices in south-eastern and parts of south-western England was sufficient in my opinion to "release" purchasing power for the purchase of manufactures' (John, 1967, pp. 191, 193). In other words, the so-called depression of 1730−50 made available surplus income for spending on an improved diet and additional luxury items. Favourable conditions for the consumer are sometimes thought to have continued through the period 1750−80 (Eversley, 1967). The difficulty is to measure consumption patterns with any confidence. Strong beer consumption, for example, fell from 33 gallons per head per annum in 1688 to 21 gallons in 1751, but this was partly because spirit consumption increased nine-fold over the period 1700−50. However, the trend in meat and dairy product consumption suggests that the agriculturally depressed years 1730−50 did not have a very strong effect on consumer demand (Beckett, 1982). The current view is that the impact of the agricultural depression between 1730−50 was probably not of a substantial magnitude relative to other causes of the forthcoming Industrial Revolution (Mokyr, 1985).

In the second half of the eighteenth century the situation changed. The classical economists, particularly David Ricardo, took the view that agriculture was a potential brake on economic progress in this period because the country was not self-supporting. He and others, notably Malthus, argued that industrialization occurred in spite of the obstacles placed in its way by a conservative and sometimes hostile agrarian sector. They expected economic progress to founder on the rocks of supply constraints from agriculture. This pessimistic view may partly reflect the fact that it was formulated in the unique circumstances of the Napoleonic wars when the world economy was disrupted, but it continued to attract supporters well into the nineteenth century, at least in part because of what was seen as the poor performance of agriculture. Ostensibly this would seem to be at odds with the concept of a post-1750 Agricultural Revolution.

As we have already seen, measured simply in terms of output English agriculture appears to have been very successful over the period 1750–1850, but did rising supply keep pace with demand? Deane and Cole (1969) showed that agricultural output grew slowly to 1740 or thereabouts but then increased more rapidly, just at the point when the Agricultural Revolution was deemed to have begun in conjunction with its industrial counterpart. Unfortunately their method of calculation was probably wrong, as Deane has since acknowledged (Floud and McCloskey, 1981, p. 64). More recently Jackson has calculated that a phase of strong growth to about 1740 was followed 'by fifty years of much slower agricultural growth, perhaps even of near stagnation' (1985, p. 333). Crafts has estimated that whereas agricultural output grew at 0.6 per cent per annum over the period 1700–60 the rate fell to 0.13 per cent 1760–80 (Crafts, 1985). Overton's evidence of wheat yields in Norfolk and Suffolk supports this view, pointing to a pattern of steady increase down to 1801 with a slow-down in the rate of growth in output per acre between 1750–1830. He suggests that average yields failed to rise because more and more physically marginal land was brought into cultivation (Overton, 1986). A further reason for this relatively poor performance was that the major gains to productivity came from new cropping systems which were introduced before 1760. Consequently the main growth of productivity associated with them was before $c.1770$.

As a result, the surplus grain balances of the period before 1750 evened out by the mid-1760s, as the home population took up the slack in output, and turned into deficits by the 1770s. To prevent starvation the gap had to be bridged by imports. By the early nineteenth century about one sixth of all grain was imported from Ireland and the Continent (see figure 1). The figure rose to 22 per cent by 1841. According to Mokyr:

notwithstanding the achievements of the agricultural revolution, the supply of grains, butter, meat, and livestock available to the English population would have been smaller by at least one-sixth in the 1830s and early 1840s had it not been for imports, mainly from Ireland. (Mokyr, 1985, pp. 147–8)

Coupled with rising prices due to demand, agriculture was seen to be effectively restraining gains in real wages in the urban economy, hence the pessimism of the classical economists.

The gap which opened up between home production and overseas imports is usually taken to reflect badly on agriculture. Given the demographic revolution of the eighteenth and nineteenth centuries it is hard on the farming community to judge it as having let down the nation when rising output only just failed to meet demand from a fast expanding population. Jones has calculated that in 1800 about 90 per cent of the British population was fed by domestic agricultural production, whereas the figure a century earlier was nearer 101 per cent (Jones, 1981, p. 68). Be that as it may, imports are usually seen as taking place because the farming community could not keep pace with demand and there were real threats of famine. In fact, the available price evidence suggests that this is too alarmist, and imports may have served to keep down prices on the home market.

No fewer than 14 of the 22 grain harvests between 1793 and 1814 were deficient in varying degrees, with particularly disastrous harvests in 1795–6 and 1799–1800. Coming as they did at a time of rapid population growth, these bad seasons led to prices reaching record levels by 1810. In an effort to increase output waste lands

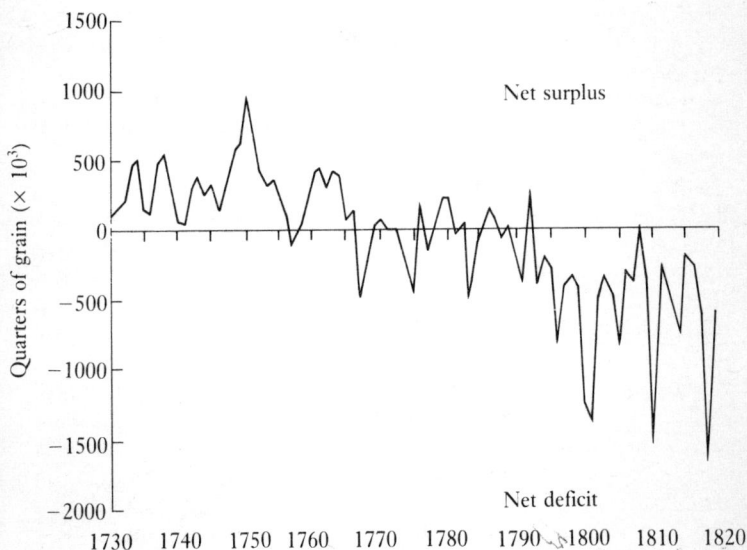

Figure 1 Time trend of the 'wheat balance', 1731–1819 (i.e. the net import–export position).
Source: M. Turner, *English Parliamentary Enclosure*. (Folkestone: Dawson, 1980).

and lands of marginal quality were enclosed. As a result, in the early years of the nineteenth century growth rates began to exceed those of the pre-1760 period (Jackson, 1985). Crafts has estimated growth at 0.75 per cent between 1780 and 1801 and 1.18 per cent between 1801 and 1831 (Crafts, 1985). After 1812 prices went into reverse. Despite the Corn Laws, which served to keep cereal prices above those prevailing in much of northern Europe, by the 1840s grain prices were little higher than in the 1770s. The result was deflation in the prices of agricultural goods, led by wheat. This suggests that output rose in such a way as to keep prices relatively low compared to total population, since there is no statistically significant evidence of changing consumer preference before 1850.

The classical economists were perhaps not as far from the truth as has sometimes been thought. Agriculture did have difficulty raising the level of supply to keep pace with demand – although not as much difficulty as some contemporaries anticipated – and it was only the import of cheap food from the Americas and Australasia after 1870 that brought down prices and pushed up the living standards of the mass of the population, both rural and urban. For all the impressive innovations, the significant enclosures and the rising productivity of land, it is at least arguable that the capacity of British agriculture to bring about change 'does not seem all that impressive' when measured in terms of the additional supply required to maintain stable prices (O'Brien, 1977).

Based simply on food supply this picture of agriculture as a prospective drain on industrial development is not an especially positive one, but what of other areas in which the two sectors interacted? We cannot look in any detail at all the links, but there can be little doubt that the Industrial Revolution required strong support from agriculture; indeed, there must have been a sizeable *net* transfer of real resources from agriculture into industry and services. Two examples can be given of how this operated. First, English landlords invested part of the income they derived from rents paid on their estates in certain types of industry, transport improvements and urban building (Beckett, 1986). Second, rising agricultural productivity made it possible for labour to be transferred to industry. The numbers employed in agriculture grew between 1700 and 1850, but demographic change ensured that there was always a steady trickle of individuals migrating to towns, a process encouraged by the new farming methods which made it possible for the agricultural sector to raise output per worker by making fuller use of a workforce which had previously been underemployed for much of the year. Accelerating population growth and a declining labour requirement allowed a significant reshuffling of the labour force in the direction of industry (Mokyr, 1985).

The process was not of course all one way. Landowners invested in road and canal improvements which benefited industry but also helped to widen the market in which agricultural products could be sold. The urban markets themselves were regarded by French commentators as offering a stimulant to English farmers to engage in capitalist production on a commercial scale which was entirely lacking across the Channel (Mingay, 1977). Industrial innovation was transferred back into agriculture through machinery. The relationship between industry and agriculture was a strong one, but the old view of the one as riding on the back of the other is no longer accepted. Today agriculture is not seen as the trigger setting off the Industrial Revolution. Several historians have emphasized instead the contribution of overseas trade; yet others, conversely, have argued that overseas trade, far from being a stimulant to the economy, grew as a result of stimulation from industrialization. Possibly that should put agriculture back centre-stage, but not everyone agrees. O'Brien, for example, has recently argued that the Industrial Revolution was 'quintessentially industrial, commercial and urban' with agriculture firmly in the background (O'Brien, 1985, p. 786).

In this welter of figures and arguments it is easy to overlook the real achievements of the farming community through the eighteenth and nineteenth centuries in raising output to feed a growing population. That there was a shortfall between supply and demand is not in question, nor that it originated in the second half of the eighteenth century just as − according to the traditional arguments − the Agricultural Revolution was beginning. The situation could, however, have been much worse. Had agriculture failed to respond positively to rising population one of two results must have followed; either the new towns of industrial England would have had to be fed from foreign imports thereby ensuring that much larger quantities of money would have been deflected from the economy, and severely retarding the pace of industrialization; or population growth itself would have been stunted. The real achievement of agriculture in this period was in fact a very considerable one. Agriculture may not have been the engine driving the Industrial Revolution, but it played its part in ensuring optimum conditions and the overall achievement was nothing short of revolutionary.

Epilogue

Little now remains of Ernle's views, and the old certainties about the Agricultural Revolution seem misplaced. Now we know more about regional differences in the pattern of innovation and enclosure, and about the growth of productivity, historians have traced a series of revolutions in agriculture from the mid-sixteenth century to the late nineteenth century. Over the century or so before 1760 new crops were introduced, and productivity increased; the period 1760 onwards saw the spread and application of the new techniques and with them a steady growth of output; after 1830 yields per acre increased greatly, and English agriculture was recognizably capitalist in structure even though many farms remained small and machinery was only slowly adopted. How these trends are interpreted is a matter of opinion. Was the Agricultural Revolution all over by 1767, as Kerridge claimed, just beginning in 1760 as others argue, or was it a longer-term movement with a second revolution after about 1815? Or should we just accept that the term 'revolution' is simply inappropriate in this context? Maybe the answer is to view what happened in agriculture in the same way that historians now see the progress of industry. Here, also, the emphasis is on a period of slow growth down to 1830 and more rapid development thereafter. The parallel with agriculture is striking. The late eighteenth and early nineteenth centuries represent a period of innovation and experiment, with strong regional differences. Norfolk led agricultural change just as Lancashire was in the forefront of industrial change. After 1830, as in industry, the whole agricultural economy showed signs of improving output and productivity to levels which had not previously been achieved but which were sustained. Perhaps the link between the two revolutions was a strong one after all, even if it took a rather different form from the one we have usually assumed.

References and Further Reading

An asterisk indicates books and articles particularly relevant for further reading.

Allen, R. C. 1982: The efficiency and distributional consequences of eighteenth-century enclosures. *Economic Journal*, XCII, 937–53.

Allen, R. C. 1986a: Enclosure, depopulation, and inequality in the south Midlands, 1377–1801. University of British Columbia Discussion Paper, 86–36.

Allen, R. C. 1986b: Enclosure, capitalist agriculture and the growth of corn yields in early modern England. University of British Columbia Discussion Paper, 86–39.

Allen, R. C. 1988: The growth of labour productivity in early modern English agriculture. *Explorations in Economic History*, 25, 117–46.

Allen, R. C. and Ó Gráda, C. 1988: On the road again with Arthur Young: English, Irish and French agriculture during the industrial revolution. *Journal of Economic History*, XLVIII, 93–116.

Allen, W. 1736: *Ways and Means of Raising the Value of Land.* London.

Armstrong, A. 1988: *Farmworkers: a Social and Economic History 1770–1980.* London: Batsford.

Ashton, T. S. 1972: *An Economic History of England: the Eighteenth Century.* London: Methuen.

Banks, S. J. 1988: Nineteenth-century scandal or twentieth-century model? A new look at 'open' and 'close' parishes. *Economic History Review*, 41, 51–73.

Batchelor, T. 1808: *General View ... of the County of Bedford.* London.

Beastall, T. W. 1978: *The Agricultural Revolution in Lincolnshire.* Lincoln: History of Lincolnshire Committee.

Beckett, J. V. 1977: English landownership in the later seventeenth and eighteenth centuries: the debate and the problems. *Economic History Review*, 33, 567–81.

Beckett, J. V. 1982: Regional variation and the agricultural depression 1730–1750. *Economic History Review*, 35, 35–51.

Beckett, J. V. 1983a: The debate over farm sizes in eighteenth and nineteenth century England. *Agricultural History*, 57, 308–25.

Beckett, J. V. 1983b: Absentee landownership in the later seventeenth and early eighteenth centuries: the example of Cumbria. *Northern History*, 19, 87–107.

Beckett, J. V. 1984: The pattern of landownership in England and Wales, 1660–1880. *Economic History Review*, 37, 1–22.

Beckett, J. V. 1986: *The Aristocracy in England, 1660–1914.* Oxford: Basil Blackwell.

Beckett, J. V. 1989: *A History of Laxton: England's last open field village.* Oxford: Basil Blackwell.

Beckett, J. V. n.d.: The decline of the small landowner in England and Wales, 1660–1900. Forthcoming.

Bourne, G. 1912: *Change in the Village.* New York: Kelley (1969 reprint).

Bowie, G. G. S. 1987: Watermeadows in Wessex: a re-evaluation of the period 1640–1750. *Agricultural History Review*, 35, 151–8.

Brenner, R. 1976: Agrarian class structure and economic development in pre-industrial Europe. *Past and Present*, 70, 30–75.

Brenner, R. 1982: The agrarian roots of European capitalism. *Past and Present*, 97, 16–113.

Broad, J. 1980: Alternate husbandry and permanent pasture in the Midlands, 1650–1800. *Agricultural History Review*, 28, 77–89.

Brown, J. and Beecham, H. A. 1989: Arable farming: farming practices. In G. E. Mingay (ed.), *The Agrarian History of England and Wales, vol. 6: 1750–1850.* Cambridge University Press, 276–96.

Butlin, R. A. 1979: The enclosure of open fields and extinction of common rights in England c.1600–1750: a review. In H. S. A. Fox and R. A. Butlin (eds), *Change in the Countryside; Essays in Rural England, 1500–1900.* London: Institute of British Geographers Special Publications no. 10, 65–82.

Caird, J. 1852: *English Agriculture in 1850–1.* London.

Caird, J. 1878: *The Landed Interest and the Supply of Food.* London.

Chambers, J. D. 1953: Enclosure and labour supply in the industrial revolution. *Economic History Review*, 5, 319–43.

* Chambers, J. D. and Mingay, G. E. 1966: *The Agricultural Revolution, 1750–1880.* London: Batsford.

Chapman, J. 1984: The chronology of English enclosure. *Economic History Review*, 37, 557–9.

Chapman, J. 1987: The extent and nature of parliamentary enclosure. *Agricultural History Review*, 35, 25–35.

Chartres, J. A. 1984–5: The marketing of agricultural produce. In J. Thirsk (ed.), *The Agrarian History of England and Wales, vol. 5: 1640–1750.* 2 vols. Cambridge University Press, 406–502.

Chorley, G. P. H. 1981: The agricultural revolution in northern Europe, 1750–1880: nitrogen, legumes, and crop productivity. *Economic History Review*, 34, 71–93.

Clapham, J. H. 1939: *An Economic History of Modern Britain.* Cambridge University Press, vol. 2.

Clark, G. 1987: Productivity growth without technical change in European agriculture before 1850. *Journal of Economic History*, XLVII, 419–32.

Coleman, D. C. 1977: *The Economy of England, 1450–1750.* Oxford: Oxford University Press.

Collins, E. J. T. 1969: Harvest technology and labour supply in Britain, 1790–1870. *Economic History Review*, 22, 453–73.

Collins, E. J. T. 1976: Migrant labour in British agriculture in the nineteenth century. *Economic History Review*, 29, 38–59.

Collins, E. J. T. 1987: The rationality of 'surplus' agricultural labour: mechanization in English agriculture in the nineteenth century. *Agricultural History Review*, 35, 36–46.

Collins, E. J. T. and Jones, E. L. 1967: Sectoral advance in English agriculture, 1850–80. *Agricultural History Review*, 15, 65–81.

Crafts, N. F. R. 1985: *British Economic Growth during the Industrial Revolution*. Oxford: Oxford University Press.

Craigie, P. G. 1887: The size and distribution of agricultural holdings in England and abroad. *Journal of the Royal Statistical Society*, 50, 86–142.

Deane, P. and Cole, W. A. 1969: *British Economic Growth, 1688–1959*. 2nd edn. Cambridge University Press.

Defoe, D. 1971: *A Tour Through the Whole Island of Great Britain*. (ed.) Pat Rogers, Harmondsworth: Penguin.

Ernle, Lord (R. E. Prothero) 1961: *English Farming Past and Present*. 6th edn. London: Heinemann.

Eversley, D. E. C. 1967: The home market and English economic growth. In E. L. Jones and G. E. Mingay (eds), *Land, Labour and Population*. London: Arnold, 206–59.

Fenoaltea, S. 1988: Transaction costs, whig history, and the common fields. *Politics and Society*, 16, 171–240.

Floud, R. and McCloskey, D. (eds), *The Economic History of Britain since 1700, vol. 1, 1700–1860*, Cambridge University Press.

General Report, 1808: *General Report on Enclosures, drawn up by order of the Board of Agriculture*. London.

Glennie, P. 1988a: Continuity and change in Hertfordshire agriculture 1550–1700: I – patterns of agricultural production. *Agricultural History Review*, 36, 55–76.

Glennie, P. 1988b: Continuity and change in Hertfordshire agriculture 1550–1700: II – trends in crop yields and their determinants. *Agricultural History Review*, 36, 145–61.

Goddard, N. 1989: Agricultural literature and societies. In G. E. Mingay (ed.), *The Agrarian History of England and Wales, vol. 6: 1750–1850*, Cambridge University Press, 361–83.

Gonner, E. C. K. 1912: *Common Land and Inclosure*. London: Macmillan.

Grigg, D. B. 1963: Small and large farms in England and Wales. *Geography*, 48, 268–79.

Grigg, D. B. 1966: *The Agricultural Revolution in South Lincolnshire*. Cambridge University Press.

Grigg, D. B. 1987: Farm size in England and Wales from early Victorian times to the present. *Agricultural History Review*, 35, 179–90.

Habakkuk, H. J. 1939–40: English landownership, 1680–1740. *Economic History Review*, x, 2–17.

Habakkuk, H. J. 1965: La disparition du paysan anglais. *Annales E. S. C.,* 29, 649−63.

Hammond, J. L. and Barbara, 1911: *The Village Labourer 1760−1832.* London: Longmans, Green and Co.

Harris, A. 1961: *The Rural Landscape of the East Riding of Yorkshire, 1700− 1850.* Oxford: Oxford University Press.

Havinden, M. A. 1961: Agricultural progress in open-field Oxfordshire. *Agricultural History Review,* 9, 73−83.

Havinden, M. A. 1974: Lime as a means of agricultural improvement: the Devon example. In C. W. Chalklin and M. A. Havinden (eds), *Rural Change and Urban Growth 1500−1800.* London: Longman, 104−34.

Healy, M. J. R. and Jones, E. L. 1962: Wheat yields in England, 1815− 59. *Journal of the Royal Statistical Society,* 125, 574−9.

Historical Manuscripts Commission, 1934: *Hastings MSS,* III.

Hobsbawm, E. J. and Rude, G. 1969: *Captain Swing.* London: Lawrence and Wishart.

Holderness, B. A. 1972: The agricultural activities of the Massingberds of South Ormsby, Lincolnshire, 1638−1750. *Midland History,* 1, 15−25.

Holderness, B. A. 1988: Agriculture 1770−1860. In C. H. Feinstein and S. Pollard (eds), *Studies in Capital Formation in the United Kingdom, 1750−1920.* Oxford: Clarendon Press, 9−34.

Holderness, B. A. 1989: Prices, productivity and output. In G. E. Mingay (ed.), *The Agrarian History of England and Wales, vol. 6: 1750−1850.* Cambridge University Press, 84−189.

Hoskins, W. G. 1957: *The Midland Peasant.* London: Macmillan.

Hoskins, W. G. 1968: Harvest fluctuations and English economic history. *Agricultural History Review,* 16, 15−31.

Hunt, H. G. 1958−9: Landownership and enclosure, 1750−1830. *Economic History Review,* 11, 497−505.

* Jackson, R. V. 1985: Growth and deceleration in English agriculture, 1660−1790. *Economic History Review,* 38, 333−51.

John, A. H. 1959: Aspects of English economic growth in the first half of the eighteenth century. *Economica,* 28, 176−90.

John, A. H. 1960: The course of agricultural change 1660−1760. In L. S. Pressnell (ed.), *Studies in the Industrial Revolution.* London: Athlone Press, 125−55.

John, A. H. 1967: Agricultural productivity and economic growth in England, 1700−60. In E. L. Jones (ed.), *Agriculture and Economic Growth in England, 1650−1815.* London: Methuen, 172−93.

Johnson, A. H. 1909: *The Decline of the Small Landowner.* Oxford: Oxford University Press.

Jones, E. L. 1965: Agriculture and economic growth in England, 1660− 1750: agricultural change. *Journal of Economic History,* 25, 1−18.

* Jones, E. L. 1967: *Agriculture and Economic Growth in England, 1650− 1815.* London: Methuen.

* Jones, E. L. 1968: *The Development of English Agriculture 1815−1873.*

London: Macmillan.

* Jones, E. L. 1974: *Agriculture and the Industrial Revolution*. Oxford: Basil Blackwell.

* Jones, E. L. 1981: Agriculture, 1700−80. In R. Floud and D. McCloskey (eds), *The Economic History of Britain since 1700, vol. 1, 1700−1860*. Cambridge University Press, 66−86.

Kent, N. 1775: *Hints to Gentlemen of Landed Property*. London.

* Kerridge, E. 1967: *The Agricultural Revolution*. London: Allen and Unwin.

Kussmaul, A. 1981: *Servants in Husbandry in Early Modern England*. Cambridge University Press.

Leonard, E. M. 1905: The enclosure of common fields in the seventeenth century. *Transactions of the Royal Historical Society*, 19, 101−46.

Levy, H. 1911: *Large and Small Holdings: a Study of English Agricultural Economics*. Cambridge University Press.

Lindert, P. and Williamson, J. G. 1982: Revising England's social tables, 1688−1812. *Explorations in Economic History*, 19, 385−408.

McCloskey, D. N. 1975: The economics of enclosure: a market analysis. In W. N. Parker and E. L. Jones (eds), *European Peasants and their Markets*, Princeton University Press, 73−119.

Macdonald, S. 1979a: The diffusion of knowledge among Northumberland farmers, 1780−1815. *Agricultural History Review*, 27, 30−9.

Macdonald, S. 1979b: The role of the individual in agricultural change: the example of George Culley of Fenton, Northumberland. In H. S. A. Fox and R. A. Butlin (eds), *Change in the Countryside; Essays in Rural England, 1500−1900*. London: Institute of British Geographers Special Publications no. 10, 5−21.

Marshall, T. H. 1929−30: Jethro Tull and the 'new husbandry' of the eighteenth century. *Economic History Review*, 2, 41−60.

Marshall, W. 1790: *The Rural Economy of the Midland Counties*. 2 vols. London.

Marshall, W. 1818: *Review and Abstract of the County Reports to the Board of Agriculture*. 4 vols. York.

Martin, J. M. 1979a: The small landowner and parliamentary enclosure in Warwickshire. *Economic History Review*, 32, 328−43.

Martin, J. M. 1979b: Members of parliament and enclosure: a reconsideration. *Agricultural History Review*, 27, 101−9.

Martin, J. M. 1984: Village traders and the emergence of a proletariat in south Warwickshire, 1750−1851. *Agricultural History Review*, 32, 179−88.

Marx, K. 1867: *Capital*, II. Everyman edn, 1930.

Mingay, G. E. 1961−2: The size of farms in the eighteenth century. *Economic History Review*, 14, 469−88.

Mingay, G. E. 1968: *Enclosure and the Small Landowner in the Age of the Industrial Revolution*. London: Macmillan.

Mingay, G. E. 1977: *The Agricultural Revolution: Changes in Agriculture 1650−1880*. London: Adam and Charles Black.

* Mingay, G. E. (ed.) 1989: *The Agrarian History of England and Wales, vol.*

6: *1750–1850*. Cambridge University Press.

Mokyr, J. (ed.) 1985: *The Economics of the Industrial Revolution*. London: George Allen and Unwin.

Molland, R. 1959: Agriculture, 1793–1870. In Victoria County History, *Wiltshire*, IV. Oxford: Oxford University Press, 65–91.

Neeson, J. M. 1984: The opponents of enclosure in eighteenth-century Northamptonshire. *Past and Present*, 105, 114–39.

* O'Brien, P. K. 1977: Agriculture and the industrial revolution. *Economic History Review*, 30, 166–81.

* O'Brien, P. K. 1985: Agriculture and the home market for English industry 1660–1820, *English Historical Review*, 100, 773–800.

* O'Brien, P. K. and Keyder, C. 1978: *Economic Growth in Britain and France, 1780–1914*. London: Allen and Unwin.

* Ó Gráda, C. 1981: Agricultural decline, 1860–1914. In R. Floud and D. McCloskey (eds), *The Economic History of Britain since 1700, vol. 2, 1860–1970s*. Cambridge University Press, 175–97.

Ormrod, D. 1985: *English Grain Exports and the Structure of Agrarian Capitalism 1700–1760*. Hull University Press.

Orwin, C. S. and Whetham, E. H. 1964: *History of British Agriculture, 1846–1914*. London: Longmans, Green and Co.

Outhwaite, R. B. 1986: Progress and backwardness in English agriculture, 1500–1650. *Economic History Review*, 39, 1–18.

Overton, M. 1979: Estimating crop yields from probate inventories: an example from East Anglia, 1585–1735. *Journal of Economic History*, 39, 363–78.

Overton, M. 1985: The diffusion of agricultural innovation in early modern England: turnips and clover in Norfolk and Suffolk, 1580–1740. *Transactions of the Institute of British Geographers*, new series, 10, 205–21.

* Overton, M. 1986: Agricultural revolution?: England, 1540–1850. *ReFRESH*, 3, 1–4.

Overton, M. et al. 1983: *Agricultural History: Papers presented to the Economic History Society Conference*. Canterbury: Economic History Society

Parker, R. A. C. 1975: *Coke of Norfolk: A Financial and Agricultural Study 1707–1842*. Oxford: Clarendon Press.

Perkins, J. A. 1975: Tenure, tenant right, and agricultural progress in Lindsey, 1780–1850. *Agricultural History Review*, 23, 1–22.

Perry, P. J. 1974: *British Farming in the Great Depression, 1870–1914*. Newton Abbot: David and Charles.

Phillips, A. D. M. 1969: Underdraining and the English claylands, 1850–80: a review. *Agricultural History Review*, 17, 44–55.

Phillips, A. D. M. 1975: Underdraining and agricultural investment in the Midlands in the mid-nineteenth century. In A. D. M. Phillips and B. J. Turton, (eds), *Environment, Man and Economic Change*. London: Longman, 253–74.

Purdum, J. L. 1978: Profitably and timing of parliamentary land enclosures. *Explorations in Economic History*, 15, 313–26.

Reed, Michael 1984: Enclosure in north Buckinghamshire, 1500–1750. *Agricultural History Review*, 32, 133–44.

Reed, M. 1984: The peasantry of nineteenth-century England: a neglected class? *History Workshop*, 18, 53–76.

Richards, E. 1973: *Leviathan of Wealth*. London: Routledge.

Riches, N. 1937: *The Agricultural Revolution in Norfolk*. Chapel Hill: University of North Carolina.

Russell, N. 1986: *Like Engend'ring Like: Heredity and Animal Breeding in Early Modern England*. Cambridge University Press.

Saville, J. 1969: Primitive accumulation and early industrialization in Britain. *Socialist Register*, 6, 247–71.

Scarfe, N. (ed.) 1988: *A Frenchman's Year in Suffolk*. Woodbridge: Boydell Press.

Slater, G. 1907: *The English Peasantry and the Enclosure of Common Fields*. London: Constable.

Smith, D. J. 1984: *Discovering Horse-drawn Farm Machinery*. Aylesbury: Shire Publications.

Snell, K. D. M. 1985: *Annals of the Labouring Poor: Social Change and Agrarian England, 1660–1900*. Cambridge University Press.

Sturgess, R. W. 1966: The agricultural revolution on the English clays. *Agricultural History Review*, 14, 104–21.

Sturgess, R. W. 1967: The agricultural revolution on the English clays: a rejoinder. *Agricultural History Review*, 15, 82–7.

Sullivan, R. J. 1985: The timing and pattern of technological developments in English agriculture, 1611–1850. *Journal of Economic History*, XLV, 305–14.

Tate, W. E. 1967: *The English Village Community and the Enclosure Movement*. London: Gollancz.

* Tate, W. E. 1978: *A Domesday of English Enclosure Acts and Awards*. University of Reading.

Taylor, C. 1975: *Fields in the English Landscape*. London: Dent.

Thirsk, J. 1954: Agrarian history, 1540–1950. In Victoria County History, *Leicestershire*. Oxford: Oxford University Press, II, 199–264.

* Thirsk, J. (ed.) 1967: *The Agrarian History of England and Wales, vol. 4: 1500–1640*. Cambridge University Press.

Thirsk, J. 1974: New crops and their diffusion: tobacco-growing in seventeenth-century England. In C. W. Chalklin and M. A. Havinden (eds), *Rural Change and Urban Growth 1500–1800*. London: Longman, 76–103.

Thirsk, J. 1976: *The Restoration*. London: Longman.

* Thirsk, J. (ed.) 1984–5: *The Agrarian History of England and Wales, vol. 5: 1640–1750*. 2 vols, Cambridge University Press.

* Thirsk, J. 1987: *England's Agricultural Regions and Agrarian History, 1500–1750*. London: Macmillan.

Thompson, F. M. L. 1963: *English Landed Society in the Nineteenth Century*. London: Routledge and Kegan Paul.

* Thompson, F. M. L. 1968: The second agricultural revolution, 1815–

1880. *Economic History Review*, 21, 62—77.

Toynbee, A. 1919: *Lectures on the Industrial Revolution of the Eighteenth Century in England*. London: Longmans, Green and Co.

Trow-Smith, R. 1957: *A History of British Livestock Husbandry to 1700*. London: Routledge and Kegan Paul.

Trow-Smith, R. 1959: *A History of British Livestock Husbandry, 1700—1900*. London: Routledge and Kegan Paul.

Turner, M. E. 1975: Parliamentary enclosure and landownership change in Buckinghamshire. *Economic History Review*, 28, 565—81.

* Turner, M. E. 1980: *English Parliamentary Enclosure*. Folkestone: Dawson.

* Turner, M. E. 1982: Agricultural productivity in England in the eighteenth century. *Economic History Review*, 35, 489—510.

Turner, M. E. 1984a: *Enclosures in Britain 1750—1830*. London: Macmillan.

Turner, M. E. 1984b: Agricultural productivity in eighteenth century England: further strains of speculation. *Economic History Review*, 37, 252—7.

Turner, M. E. 1984c: The landscape of parliamentary enclosure. In M. Reed (ed.), *Discovering Past Landscapes*. London: Croom Helm, 132—66.

* Turner, M. E. 1986a: Parliamentary enclosures: gains and costs. *ReFRESH*, 3, 5—8.

Turner, M. E. 1986b: English open fields and enclosures: retardation or productivity improvements? *Journal of Economic History*, 46, 669—92.

Turner, M. E. 1988: Economic protest and rural society: opposition to parliamentary enclosure in Buckinghamshire. *Southern History*, 10, 94—128.

Walton, J. R. 1976: Aspects of agrarian change in Oxfordshire, 1750—1880. University of Oxford D. Phil thesis. Unpublished.

Walton, J. R. 1986: Pedigree and the national cattle herd, *Circa*. 1750—1950. *Agricultural History Review*, 34, 149—70.

* Woodward, D. M. 1971: Agricultural revolution in England 1500—1900: a survey. *The Local Historian*, 9, 323—33.

Woodward, D. M. 1977: A comparative study of the Irish and Scottish livestock trades in the seventeenth century. In L. M. Cullen and T. C. Smout (eds), *Comparative Aspects of Scottish and Irish Economic and Social History, 1600—1900*. Edinburgh: John Donald.

Wordie, J. R. 1974: Social change on the Leveson-Gower estates, 1714—1832. *Economic History Review*, 27, 593—609.

* Wordie, J. R. 1983: The chronology of English enclosure, 1500—1914. *Economic History Review*, 36, 483—505.

Wright, T. 1795: *A Short Address to the Public on the Monopoly of Small Farms*. London.

Wrigley, E. A. 1985: Urban growth and agricultural change: England and the Continent in the early modern period. *Journal of Interdisciplinary History*, XV, 683—728.

* Yelling, J. A. 1977: *Common Field and Enclosure in England 1450—1850*. London: Macmillan.

Young, A. 1767: *The Farmer's Letters to the People of England*. London.

Index